Metagenomics: Techniques, Applications, Challenges and Opportunities

Reena Singh Chopra • Chirag Chopra •
Neeta Raj Sharma
Editors

Metagenomics: Techniques, Applications, Challenges and Opportunities

Springer

Editors
Reena Singh Chopra
School of Bioengineering and Biosciences
Lovely Professional University
Jalandhar, Punjab, India

Chirag Chopra
School of Bioengineering and Biosciences
Lovely Professional University
Jalandhar, Punjab, India

Neeta Raj Sharma
School of Bioengineering and Biosciences
Lovely Professional University
Jalandhar, Punjab, India

ISBN 978-981-15-6528-1 ISBN 978-981-15-6529-8 (eBook)
https://doi.org/10.1007/978-981-15-6529-8

This Springer imprint is published by the registered company Springer Nature Singapore Pte Ltd.
The registered company address is: 152 Beach Road, #21-01/04 Gateway East, Singapore 189721, Singapore

Contents

About the Editors

Reena Singh Chopra is an Assistant Professor at the School of Bioengineering and Biosciences, Lovely Professional University. She holds a Ph.D. from Shri Mata Vaishno Devi University, India. She has received IARDO and RACE-Bangkok Awards for Best Teacher (University Level). Her research focuses on metagenomics, microbial diversity, directed evolution, and mutagenesis for improving catalytic activity of microbial enzymes. She is currently exploring metagenomics for novel hydrolases and the production of bioactive molecules from myxobacteria. She has published ten papers and book chapters and is on the editorial boards of six journals. She is a member of the Indian Science Congress Association and a founding member of The Society of Biologists, Jammu and Kashmir, India.

Chirag Chopra is an Assistant Professor at the School of Bioengineering and Biosciences, Lovely Professional University. He holds a Bachelor of Technology in Industrial Biotechnology from Shri Mata Vaishno Devi University and an M.S. (Research) from the Indian Institute of Technology Madras. He has published ten papers in journals including Oncogene, Science Bulletin, and f1000Research and a book chapter. He is on the editorial board of five journals. He has received IARDO and RACE-Bangkok Young Scientist Awards. His research focuses on targeting key oncogenic signaling proteins and T-cell checkpoint in cancer through plant-based natural compounds. He is currently working on microbial remediation of petroleum hydrocarbons. He is a member of the Indian Science Congress Association and a founding member of The Society of Biologists.

Neeta Raj Sharma is the Additional Dean of the School of Bioengineering and Biosciences at Lovely Professional University, Phagwara, Punjab. Her research focuses on the fields of microbial biochemistry and biotechnology, wealth from waste and active molecules from medicinal plants. She has received research grants from the European Commission, IC-IMPACTS-DBT, and Punjab State Council for Science and Technology. She is a fellow member of the Association of Biotechnology and Pharmacy, Indian Science Congress Association, Association of

Microbiologists of India, and Indo-US Collaboration of Engineering Education and a life member of the Association for the Promotion of DNA Fingerprinting and Other DNA Technologies, Indian Society of Agricultural Biochemists. She has published over 55 papers in leading journals and two books and holds 20 patents.

Part I

Introduction to Metagenomics

The New Science of Metagenomics: Revealing the Secrets of Microbial Physiology

1

Saurabh Singh, Harpreet Singh, Biswaranjan Rout,
Raja Babu Mani Tripathi, Chirag Chopra, and Reena Singh Chopra

Abstract

The role of microorganisms is well-established in regulating the nature's activities and the applications in different industries. Whole-cells as well as microbial products such as enzyme, secondary metabolites and peptides are used in bioremediation, fermentation, pharmaceutics, food, textile industries, among others. The great plate anomaly advocates that around 99% of the microorganisms in the environment are unculturable. The corollary to this statement is that the applications mentioned above arise out of the culturable fraction only, so it is only unimaginable what the other 99% holds for us. Metagenomics provides the necessary tools for exploring the diversity of the unculturable microbes and for bioprospecting the novel genes for different applications. This is possible due to the advent of modern molecular techniques and DNA sequencing and data analytics. The chapter highlights the salient aspects of the molecular methods used in metagenomic analyses and an overview of the applications of metagenomics to acquaint the readers about the new science of metagenomics.

Keywords

High-throughput sequencing · Metagenomics · Microbial communities · Molecular markers

S. Singh · R. B. M. Tripathi · C. Chopra · R. S. Chopra (✉)
School of Bioengineering and Biosciences, Lovely Professional University, Phagwara, Punjab, India
e-mail: chirag.18298@lpu.co.in; reena.19408@lpu.co.in

H. Singh
Department of Bioinformatics, Hans Raj Mahila Maha Vidyalaya, Jalandhar, Punjab, India

B. Rout
Centre for Biotechnology, Siksha 'O' Anusandhan University, Bhubaneswar, Odisha, India

R. S. Chopra et al. (eds.), *Metagenomics: Techniques, Applications, Challenges and Opportunities*, https://doi.org/10.1007/978-981-15-6529-8_1

1.1 Introduction

Microorganisms constitute two-thirds of the Earth's biological diversity. The seemingly endless capacity of microbes touches almost every process in the living world. Soil is a house to the most complex of these communities, which have developed synergistic relationships with it. A typical example is nitrogen-fixing bacteria which converts the nitrogen from unusable to the readily usable form by plants. Many native soil microbes operate on decaying plants and animals to recycle nutrients while others convert ions into bioavailable nutrients for plants. Microbes are known to create foods or add value to them through several natural processes, including fermentation making them useful for human consumption.

Furthermore, microbes enable innovations and technologies that can improve human life and society. The marine microorganisms globally regulate the biogeochemical cycles through several (bio)chemical transformations. They also control the energy flux in the oceans, atmosphere and can influence climate change. Thus, overall the marine microbiota makes the planet habitable and sustain other life forms (Falkowski et al. 1998).

It is surprising to observe that on one hand, microbial organisms are essential players in preserving the environmental stability and the health of individuals and on the other hand, they can cope with extremities of temperature, pressure and pH levels to survive, where no other organisms can thrive. This ability of microbes to survive under extreme environment has developed due to their smart strategies for survival. Their genomes have gone through countless biochemical transformations, to make their countless generations of their communities to adapt to enormous environmental changes for the billions of years (Sleator et al. 2008). This diversity presents a bountiful of the genetic and biological pool that may be prospected to discover novel products, including novel genes, metabolic pathways and their products (Cowan et al. 2005). However, the majority of this genetic and functional diversity is still unexploited. Around 4×10^{30} to 6×10^{30} estimated prokaryotic cells are yet to be characterized which represent the most significant proportion of individual gut microbiota, comprising 10^6 to 10^8 separate genospecies (Turnbaugh and Gordon 2009; Sleator et al. 2008). The study of ubiquitous microorganisms is especially challenging when the task is to discover novel microorganisms or novel applications of known microorganisms. For such studies, in-depth knowledge of the interaction of microorganisms with the environment (Escobar-Zepeda et al. 2015).

Moreover, in the present scenario marked by unprecedented and dramatic global changes, understanding the dynamic role of microbial communities has become more challenging. Conventionally, the microorganisms are studied in the form of pure cultures, which requires extensive culture work. However, this approach limits one's understanding of the microbial communities; after all, the microbes function naturally in communities. Nevertheless, with the advancement in science and technology, we have developed strategies to study microbes of the complex communities in their natural environments, thus making us more capable of understanding their capabilities and specific roles. The lacuna in single-microbe culture-based studies is that such studies give us information of only ~10% of the total microbial community.

The great-plate anomaly (Staley and Konopka 1985) describes the "unculturability" of the microorganisms in an environmental sample. Therefore, the study of whole-microbial communities is preferred over single-species analysis, when the aim is to understand the community functions and their potential applications.

1.2 Microbial Community Study: A Historical Perspective

Microbial communities comprise of all such microorganisms that co-exist in an ecosystem and share the resources and the community outputs (Begon et al. 1986). The first observation of a microorganism was made in the latter half of the seventeenth century when Antonie Van Leeuwenhoek observed oral microbiota for the first time using an indigenous microscope. Since then, our understanding of microbes and their communities has gone far ahead using the current molecular techniques. However, this journey was not straight forward and has been marked by enormous contributions from some pioneers. Isolation and cultivation of these "invisible" organisms by Robert Koch, later, helped to understand the various aspects of microbial physiology (Blevins and Bronze 2010). Soon, the term invisible became obsolete with the advent of microbiological staining methods. Some noteworthy contributors were Gram's staining, Schaffer-and-Fulton staining and Ziehl-Neelsen staining. These staining methods provided valuable insights into the biochemical make-up of various microorganisms (Beveridge 2001; Blevins and Bronze 2010).

Winogradsky achieved a breakthrough by establishing formulations of the culture media which were similar to the natural growth conditions of symbiotic microbes (McFall-Ngai 2008). These growth media enabled the culture and maintenance of microorganisms in pure form. In due course of time, the concept of microbial ecology was introduced due to the work of Winogradsky. Microbial ecology involves the study of microorganisms and their functions in the communities (Ackert 2012). The culture-based approaches were applied to the studies of microbial communities including aquatic microbes as well. Apart from culturing and routine staining, the biochemical tests were also developed for identification of microbial species (Colwell et al. 1996). These techniques provided insights into the microbial world, but in the present scenario, they extract limited information, to be used for other applications. Therefore, the science of metagenomics came into being. In metagenomics, the focus is on exploring the unculturable fraction of the environmental microbial communities. The unculturable microorganisms form the majority of any environmental sample and cannot be analysed using conventional microbial culture techniques. Metagenomic researchers extract the DNA directly from the environmental sample via a direct extraction method. On the other hand, some researchers also use the indirect DNA extraction methods for DNA extraction in which the cells are isolated from the environmental sample before DNA extraction, without the need for culturing them. From the extracted metagenomic DNA, the metagenomic libraries are constructed in suitable vectors. These vectors are then transformed into a host bacterium and followed by screening either by sequencing or testing based on some physiological functions. Metagenomic libraries are constructed in such a way that each part of DNA is at least represented 2–3 times. Small DNA fragments (2–3 kb) provide better coverage as compared to the larger fragments. Small fragments are useful in phenotypic studies involving single genes while reconstructing metagenomes for genotypic analysis. However, large fragments are very much desirable when exploring multigene metabolic pathways. Roughly, at least 1011 genomic clones are required to sequence genomes from rare members of microbial communities (Rastogi and Sani 2011).

From the day life started in our planet 'The Mother Earth' has been loaded with harmful pollutants from multiple sources. The pollution is a global concern because natural, as well as synthetic compounds, are coming to the environment that creates many health problems to humans and the entire ecosystem. Among the various types of environmental pollution, wastewater from the industries is the most significant category of different pollution (affecting water and soil). From public domain, several reports have confirmed that companies and industries use a plethora of chemicals to process raw materials to produce consumer products (Maszenan et al. 2011; Megharaj et al. 2011). However, the quality and biodegradability of the reagents and Xeno-products may not always agree with the environment. These substances have hazardous effects on the environment, causing an imbalance in the biosphere. The industries that produce wastewater have high chemical oxygen demand, total suspended solid levels, biochemical oxygen demand, along with organic and inorganic pollutants (Techtmann and Hazen 2016). Inorganic pollutants contain heavy metals like chromium, cadmium, arsenic, lead and mercury. Organic pollutants in wastewater contain azo-dyes, pesticides and phenol derivatives (such as phenol derivatives are chlorinated phenols, polyaromatic hydrocarbons, polychlorinated biphenyls and endocrine-disrupting chemicals). Worldwide, heavy-metal pollution is increasing at an alarming rate. Contamination with toxic metals like arsenic and mercury portray a particular bioremediation peril. Mercury is a toxic heavy metal that accumulates in the environment and causes diseases like kidney failure and blood poisoning in mammals (Maszenan et al. 2011).

Microbes have the potential to eliminate inorganic and organic pollutants from industrial waste. These organisms (fungi, algae, bacteria) typically use bioremediation through biodegradation to manage wastewater. Biodegradation is the removal of very toxic complex organic contaminants into less toxic or non-toxic molecules. For persistent pollutants which do not get easily removed a multi-step procedure involving various microbial population and enzymes are used (Maszenan et al. 2011). The process widely depends on the potential of microbe's metabolic functions to remove pollutants from wastewater (Maszenan et al. 2011; Megharaj et al. 2011; Saxena et al. 2016). The process of bioremediation can be done both ex-situ and in-situ, but this process can be slow at times, going at a pace at which the pollutant's concentration keeps increasing over time. Even, adding non-native enzymes can often lead to contamination of water body (Antizar-Ladislao 2010). Therefore, to understand the microbial community composition better, the approach of metagenomics is used. It includes technologies like high-throughput sequencing that provides reliable

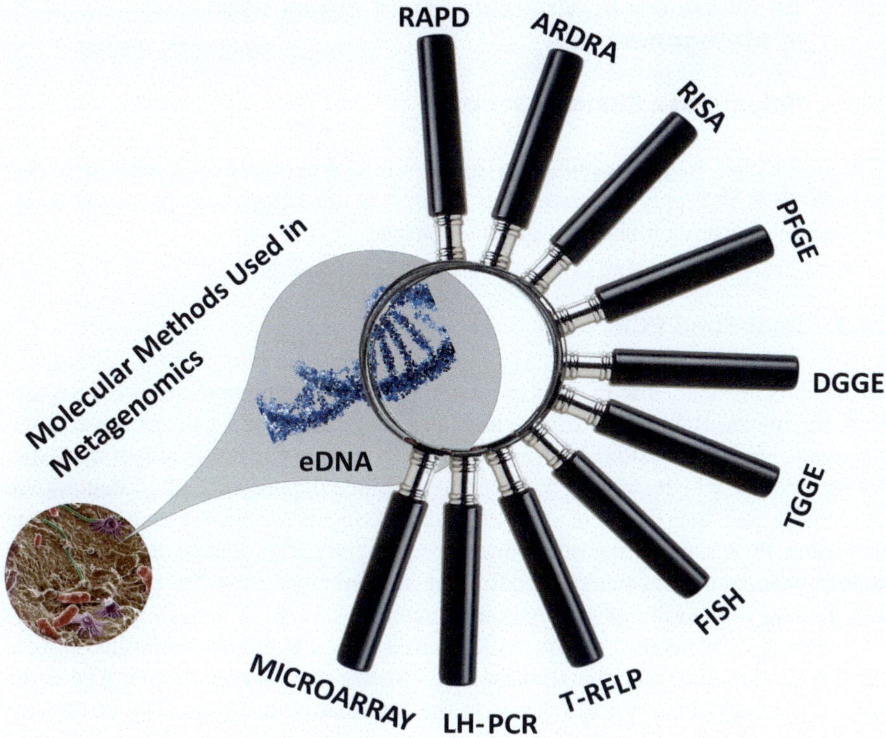

Fig. 1.1 Illustration showing the molecular markers and techniques used in metagenomics

information about the crucial genes involved in the degradation or transformation of environmental pollutants (Techtmann and Hazen 2016). Other molecular techniques can be used for monitoring the efficiency of microbial organisms in detoxification of the wastewater in the process of bioremediation (Desai et al. 2010). Various traditional methods that have been applied to determine the functionality of the microbial community in diverse environments and frequently at the contaminated ecological sites, are illustrated in Fig. 1.1. These methods include techniques like random amplified polymorphic DNA (RAPD), (amplified ribosomal DNA restriction analysis (ARDRA), rapid identification of SSRs and analysis of primers (RISA), pulsed-field gel electrophoresis (PFGE), denaturing gradient gel electrophoresis (DGGE), fluorescent in-situ hybridization (FISH), temperature gradient gel electrophoresis (TGGE), terminal restriction length polymorphism (TRFLP), length heterogeneity-polymerase chain reaction (LH-PCR), nucleic acid microarrays and on-chip technology. These techniques are explained in the latter part of the chapter.

1.3 An Overview of Molecular Techniques Used in Metagenomics

1.3.1 Polymerase Chain Reaction

PCR, a standard method applied to make multiple replicas of a section of the genome. PCR is highly accurate. It is utilized to intensify, or copy, a particular DNA segment from a mixture of genetic material.

1.3.2 Real-Time PCR

The advents in Real-Time PCR have altered the evaluation procedures of RNA and DNA fragments. Real-time PCR permits an exact estimation of the genome abundance and gene expression. It is an accurate technique for the evaluation of species abundance, which can be significant to conclude the presence of pathogens and hereditary infections (Iyer et al. 2001; Horak and Snyder 2002). Genes of interest in Real-Time PCR can be measured more prominently, reproduced accurately and quickly examined for enhanced control of assessment during the procedure and have a lesser probability of getting contamination.

Real-time PCR needs a thermocycler paired with a computer-controlled optical framework for recording the fluorescence (Miller and Barnes 1986). There are likewise contrasts between in regards to the information handling. The outflow of fluorescence creates a sign that increments in straight extent among the measure of PCR items. Fluorescence signals are recorded throughout every cycle and tell the measured intensified item. The fluorescent substances utilized can be TaqMan® or SYBR® Green.

1.3.3 Fluorescence In Situ Hybridization (FISH)

Fluorescence in situ hybridization (FISH) is a cytogenetic technique that uses complementary fluorescence-tagged probes to depict the presence or absence of a target sequence on a chromosome. Fluorescence microscopy is utilized for discovering the location of fluorescent probe binding to the chromosome. FISH conveniently locates the genes or parts of a gene on the chromosome in a single cell (Miller and Barnes 1986). This enables to study the structural, numerical and gene-level mutations in the chromosomes. Not quite the same as most different methods utilized for chromosomes study, FISH need not be performed on cells that are effectively isolating, which makes it a flexible strategy.

FISH is also used for gene mapping and identification of novel oncogenes. Also, FISH applies to detection of infectious microbes. Latest modifications in FISH technology involves techniques for improving labelling probes efficiency and also using high-resolution imaging systems for on spot visualisation of intranuclear chromosomal arrangement and the report of RNA transcription in each cell (Cui et al. 2016). These convenient, however viable, methods have revolutionized cytogenetics and have shown potential as a simple method of cancer research.

1.3.4 Host-Specific 16S rDNA Sequencing

16S rDNA sequencing techniques have been instrumental in reliably recognizing bacterial isolates and finding novel bacteria (Woo et al. 2008). The gene encoding rRNA, a part of small ribosome subunit, constitute both very uneven and preserved parts which allow people to study and differentiate and classify all organisms within the genome (Błaszczyk et al. 2011). 16S rDNA sequencing is especially crucial for specific phenotypic profiles of unusual bacteria, slow-dividing bacteria, bacteria that are highly uncultivable and culture-negative infections (Woo et al. 2008).

1.3.5 Amplified Ribosomal DNA Restriction Analysis (ARDRA)

Amplified Ribosomal DNA Restriction Analysis (ARDRA) is the modification of RFLP encoding the small (16s) ribosomal subunit of the bacteria. ARDRA has proven useful in differentiating bacterial variety at diverse taxonomic ranks, existing on the collection of retained or different regions accompanied with the digestion utilizing restriction enzymes that cut after recognising the specific four nucleotides in ribosomal genomes. Patterns seen from many restriction enzymes have been used to classify the cultivated isolates phylogenetically (Tiedje 1995).

Although ARDRA does not include details on the varieties of organisms in a mixture, helping us to determine changes at the gene level, compare various populations easily, or research the effect of environmental factors or chemicals on biocenosis resources (Gich et al. 2000; Liu et al. 1997).

1.3.6 Ribosomal Intergenic Spacer Analysis (RISA)

Ribosomal intergenic spacer analysis (RISA) is a microbiome analysis tool that gives calculates microbial range and components eliminating the discrimination forced via culture-dependent methods or research involving the creation of a small-sub-unit rRNA gene library. It was initially applied to distinct soil variety (Borneman and Triplett 1997) and, lately, for investigating microbial heterogeneity in marine environments (Acinas et al. 1999; Robleto et al. 1998). The technique includes amplification via PCR of the intergenic region's total bacterial population DNA between large (23S) and small (16S) subunit rRNA gene in rRNA operon, among nucleotide primers specified in 16S and 23S gene to conserved areas. The intergenic area 16S-23S that probably encode tRNAs based on bacterial strains shows the substantial length and sequence heterogeneity. The specific primers for these regions enable species identification as well as analysis of the closely related species (Aubel et al. 1997; Jensen et al. 1993; Maes et al. 1997; Navarro et al. 1992;

Scheinert et al. 1996). In RISA, intergenic spacer's length variability is manipulated. The PCR product (a composition of segments that group members contribute) is electrophoresed in polyacrylamide gel, and silver staining visualizes DNA. The effect is the dynamic banding pattern offering a specified community outline, among every band of DNA in the original assembly representing at least one individual (Fisher and Triplett 1999).

RISA is a rapid and straightforward approach to fingerprinting; however, the use in the microorganism's society assessment from an infected sample is limited, mostly due to the repository for the ribosomal intergenic spacer sequences are not that broad or absolute like the database for 16S selection (Spiegelman et al. 2005). The drawbacks involve the need for large segments of DNA, more time requirements, unaffected to silver staining in a few cases (Fisher and Triplett 1999).

1.3.7 Denaturing Gradient Gel Electrophoresis (DGGE)/ Temperature Gradient Gel Electrophoresis (TGGE)

In DGGE (Fischer and Lerman 1979, 1983; Myers et al. 1987) as well as TGGE, segments of the same length DNA (Rosenbaum and Riesner 1987) may be isolated with different sequences. Separation of these fragments depends on reduced mobility of partly denatured dsDNA in the polyacrylamide gel due to a linear gradient of the denaturants of DNA (a combination of formamide and urea) or a temperature gradient. The fragment melting occurs in distinct so-called melting regions: stretches of the base pairs with the same melting temperature. If a region with the lowest melting temperature touches the temperature at which it melts (Tm) at a specific location in de-naturing or temperature gradient gel, it leads to the helical transition of the partially melted molecule, and its movement is halted. In these domains, sequence variation allows the melting temperatures to vary, and the molecules with various sequences can avoid their migration at various locations in the gel (Muyzer and Smalla 1998).

DGGE/TGGE is one of microbial ecology's most common fingerprinting methods (Kirk et al. 2004). It is used in various stages for bioaugmentation approaches alone or in conjunction with cloning and sequencing. It has been applied in combination with culture-based methods to reveal phylogeny but for mainly the function and the ecological importance of various members of the complex microbial consortia corrupting pesticides (Singh et al. 2003, 2003; Bending et al. 2003; Dejonghe et al. 2003; Shi and Bending 2007; Philip et al. 2007). Many studies have used analysis of DGGE/TGGE to track the fate of decaying micro-organisms released into infected soil environment (Cunliffe and Kertesz 2006). Moreover, DGGE/TGGE (Miyasaka et al. 2006; MacNaughton et al. 1999) has been used to research potential disruptions in the microbe's soil culture, induced by the discharge of contaminant-degrading organisms during bioaugmentation or biostimulation.

1.3.8 Terminal-Restriction Fragment Length Polymorphism (T-RFLP)

Examination of PCR-amplified genes by the Terminal-RFLP (T-RFLP), a commonly executed fingerprinting method for the environmental samples. It is focused on the limitation of endonuclease digestion of the end-labelled PCR product with fluorescence. Just segments of terminal end-labelled restriction sites are identified after the examination. An electropherogram is created that displays microbial compost community profile like a series of peaks of the different height. The method is used extensively in the discovery of rare microbial ecosystems plus in research in natural habitats of bacterial, archaeal, and eukaryotic communities (Tiquia 2010).

1.3.9 Randomly Amplified Polymorphic DNA (RAPD) Analysis

Randomly amplified polymorphic DNA needs a lesser amount of the genetic material, and no background knowledge of genome is needed. The method relies on random amplification of the DNA fragments, via polymerase chain reaction, by utilizing arbitrary short primer sequences. For most species, the system detects abundant polymorphism. For certain cases, it works well and aims to become a fundamental method for genetic studies of closely associated species (Bowditch et al. 1993).

1.3.10 DNA Microchips

DNA microarray is used to measure expression levels or relative abundance of a large number of genes in a single run. This technology is an essential tool in examining gene components among the whole genome and analysing the gene expression. The usage of microarray has also been applied to 16s rRNA markers. Microarray is generally used for screening a metagenomic library with all microbial genomes. A great variety of techniques has been used during the production, incorporation of biotin-labelled nucleotides which have an affinity for fluorescently labelled streptavidin, the fusion of a modified receptive nucleotide attached to a fluorescent tag, included afterwards and an assortment of the signal intensifying techniques. The two most commonly used methods use fluorescently labelled nucleotides in the cDNA or integration of biotin labelled nucleotide in the cDNA. The labelled cDNA is then hybridised to a microarray; the cluster is washed thoroughly. The identification of signal is made via analysing fluorescence at every single point. The sample is stained post-hybridization by the fluorescently labelled streptavidin molecule. The laser incited fluorescence is regularly estimated with the microscope (scanning confocal microscope). Power of signal(s) on every point is noted as the proportion of articulation point of comparing quality (Cui et al. 2016).

1.4 Metagenomics Applications

The metagenomic approaches and the allied techniques enable bioprospecting of the unique environmental niches. Thus, metagenomic approach is invaluable for the discovery of novel ORFs, catabolic enzymes, metabolite-producing enzymes, as well as physiological aspects of bioremediation. Some of the critical applications of metagenomics are illustrated in Fig. 1.2 and reviewed in the next section of the chapter.

1.4.1 Metagenomic Discovery of Enzymes

The metagenomic approach has provided several novel catabolic enzymes for different industrial applications (Lorenz et al. 2002). These enzymes include the Cellulases, xylanases, lipases and proteases, among others. A brief review of the catabolic enzymes follows

1.4.1.1 Cellulases

Cellulases are cellulose-degrading enzymes that breakdown the cellulose present in lignocellulosic waste as well as pure cellulose, into oligosaccharides and glucose (Li et al. 2006). Cellulases have been of interest due to the diversity of applications they present. These include their use as an agent to improve the nutritional quality as well as digestibility of animal feed, fruit juice processing and baking. Recently, cellulases have been employed in de-inking (Soni et al. 2008). Various natural environments, including soil, rumen and compost, contain cellulases that can be

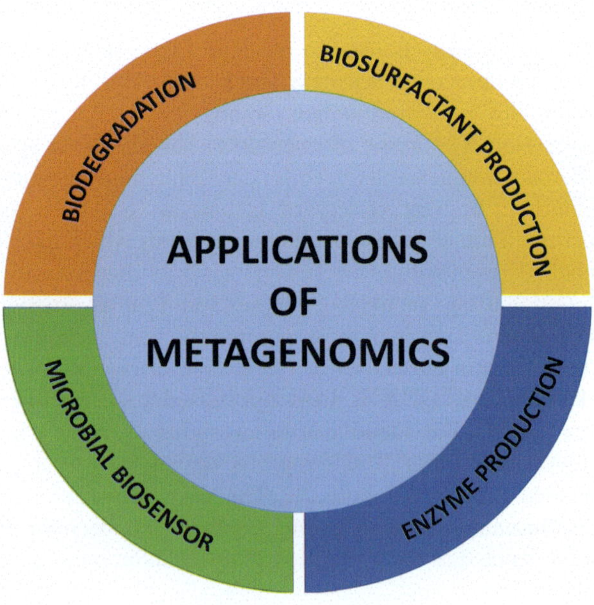

Fig. 1.2 Illustration showing some applications of metagenomics

isolated by using different metagenomic techniques. Cellulases have been isolated from a niche environment, including anaerobic digester, alkaline and saline lakes (Rees et al. 2003). These metagenomic strategies led to the isolation of many novel enzymes from a variety of environments including GH12 cellulase from rice straw compost (Yeh et al. 2013), β-glucosidases and endo-β-1,4-glucanases from the forest soil, elephant dung, cow rumen and rotten tree (Wang et al. 2009). Also, a β-glucosidase gene (unglu135B12), a member of glycoside hydrolase family 3 (GH3) (Li et al. 2014) and a novel CelEx-BR12 gene from ruminal microorganisms (Ko et al. 2013) have been reported.

1.4.1.2 Xylanases

Xylanases breakdown xylan and cleave its backbone into smaller oligosaccharides. Like cellulases, xylanases have a wide range of industrial applications including extraction and preparation of beverages (Wong et al. 1988), purification of juices (Sharma and Chand 2012), detergents (Kumar et al. 2004), generation of plant cell protoplast (Kulkarni et al. 1999), production of antimicrobial agents (Christakopoulos et al. 2003), antioxidants (Katapodis et al. 2003), surfactants (Kashyap and Subudhi 2014), among others. Xylanases have been produced by a variety of microbes living in diverse natural environments, thereby making them one of the favorite candidates for metagenomic studies. Xylanases have been produced from the metagenomic libraries as well (Ellilä et al. 2019; Knapik et al. 2019; Qian et al. 2015).

1.4.1.3 Lipases

Lipases are lipid hydrolysing enzymes, also known as triacylglycerol acyl hydrolases (EC 3.1.1.3). The lipolytic enzymes catalyze the hydrolysis of triacylglycerol to glycerol and fatty acids. They have gained special attention from the industry due to their robust nature, making them resistant to extremities of temperature, pH, organic solvents and many more. Many plants, animals and microbes contain lipases. Lipases of microbial origin are gaining particular importance due to their diverse industrial applications such as the making of oils, fats, detergents, dairy products and pharmaceuticals (Cardenas et al. 2001). The metagenomic approach is used for isolating lipases from various environmental samples including thermal (Rhee et al. 2005), saline lake (Rees et al. 2003), field soil (Henne et al. 2000), marine sediments and drinking water.

1.4.1.4 Proteases

Proteases represent one of the most important classes of enzymes having immense applications in research as well as industry. They have been present in almost all living organisms, including plants, animals and microbes. Proteases of microbial origin have gained particular interest due to their unique applications in the biotechnological and pharmaceutical industry (Jaouadi et al. 2011). Several novel proteases have been isolated using a metagenomics approach in recent times (Singh et al. 2015).

1.4.2 Metagenomics and Medicine

For a long time, microbes have been the source of many antibiotics and other medical agents and played a significant role in improving human health. However, we have almost reached a saturation point of discovering novel, useful products from microorganisms using traditional culturing methods. We need new strategies for discovering new medicinal products from microorganisms. The metagenomics approach has been promising to discover new avenues of novel antibiotics. Recently, many novel antibiotics have been discovered using this strategy including isolation of Turbomycin A and B from Soil (Gillespie et al. 2002), Didemnin B (Aplidine™) and Thiocoraline for cancer treatment (Liu et al. 2010) and red indirubin pigment possessing antibacterial activity (Lim et al. 2005).

1.4.3 Metagenomics and Biosurfactants

As chemical surfactants pose toxicity and environmental concerns, the focus is being shifted to developing biosurfactants. With the advances in biotechnology, biosurfactants are emerging as a potential replacement of chemical surfactants in industrial applications including lubrication, wetting, foaming, defoaming, emulsification, softening, fixing, among others. Also, biosurfactants have been sought out in the biomedical, food and pharmaceutical industry (Henkel et al. 2012). Metagenomic DNA libraries constructed from petroleum-contaminated sites were screened for biosurfactant production, and positive clones were identified. Morikawa and his co-workers in 1993 reported two bacteria (A-1 and B-1) which exhibited emulsified halos around their colonies on oil-L-agar plates. Very recently, a peptide biosurfactant MBSP1 extracted from the soil metagenome (da Silva Araújo et al. 2020).

1.4.4 Metagenomics and Biodegradation

There has been a continuous gathering of oil hydrocarbons as a result of oil spills and incomplete combustion of non-renewable energy sources. The seepage of these anthropogenic mixes, through industry involved in oil production and related products, leads to deposition of lots of sweet-smelling hydrocarbons, thus polluting natural systems. Microorganisms play a significant role in biogeochemical cycles, specifically in degrading a variety of carbon-based natural products. These can be therefore utilized to degrade aromatic rings such as benzene, toluene, and xylene rings, and mineralize their carbon skeleton (Alexander et al. 1994). Recently, researchers have discovered genes and their respective metabolic pathways, involved in the degradation of a variety of aromatic compounds in sludge waste using metagenomics (Silva et al. 2012). Also, bacterial populations, capable of degrading polycyclic aromatic hydrocarbon (PAH), have been identified from cold marine ecosystems (Marcos et al. 2009).

With the beginning of next-generation sequencing and analytical in-silico tools, researchers have been able to explore the uncultured microbial diversity by studying environmental DNA. Two main sequence-based strategies are being employed in metagenomics, i.e. the "gene-centric" and "genome-centric". These approaches extract information about the identity and biological requirements from the analysis of metagenomic DNA. The gene-centric method is only limited to the study of metagenomic DNA in the environment for the presence of various metabolic and functional genes as well as taxa. In contrast, the genome-centric method involves the rebuilding of a complete or partially complete genome from the metagenomic sequence data (Prosser 2015). The role of metagenomics in the process of in situ bioremediation is critical. In the process of bioremediation, the rate of decontamination is accelerated either by the addition of rate-limiting chemicals such as electron acceptors (biostimulation) or by addition of whole cells (bioaugmentation) (Czarny et al. 2019; Handelsman 2004).

1.4.5 Exploration of Microbial Sensors in the Community

Ecological pressure and anthropogenic activities such as contamination have practicably restricted or promoted the growth of particular microbial community (Darwin 2012). Based on the above fact, in many studies, the use of microorganisms as a biosensor to detect environmental alterations are reported. One such example where machine learning was used to develop a model to predict microbial community structure in groundwater contaminated with uranium and nitrates as well as oil-polluted sea samples. Groundwater samples collected from polluted aquifer from the Bear Creek watershed in Oak Ridge, Tennessee containing Uranium (ranging from undetectable to 55.3 mg/L) and nitrate (range from undetectable to 14,446 mg/L) were used to assess microbial diversity at a taxonomic level using 16S rRNA sequencing (Watson et al. 2004).

The taxonomic composition and relative richness of each taxon were analysed from the sequence data to build a computational model. This model was directed to relate microbial population structure to geochemical variables in which the microbiome structure was submitted to the algorithm for testing the match between the model's prediction and geochemical measurement. Analyses showed indication of key taxa involved in the metabolism of contaminants such as *Brevundimonas* spp., *Rhodanobacter* and *Rhodocyclaceae*. *Brevundimonas* spp. are active nitrate reducers (Kavitha et al. 2009) and *Rhodanobacter* and *Rhodocyclaceae* metabolize uranium and help in bioremediation (Green et al. 2012). The utility of this model was further validated to predict oil contamination in the marine ecosystem. Through 16S rRNA microarray (PhyloChip), seawater samples were analyzed for taxonomic composition and relative abundance of each taxon and oil concentrations were also measured based on GC/MS technology from the Gulf of Mexico during the *Deepwater Horizon* oil spills (Dubinsky et al. 2013). The presence of *Oceanospiraillales* was mostly indicated in oil-contaminated water, whereas *Pelagibacteriaceae* was indicated in non-contaminated water samples.

1.4.6 Biodegradation of Marine Oil Spills (*Deepwater Horizon* Oil Spill)

The extensive use of metagenomics for biodegradation of oil in the seawater was applied during the case of first worst marine oil spills, i.e. *Deepwater Horizon* oil spill in the USA (King et al. 2015). Nearly 4.1 million barrels of oil were accidentally released in the Gulf of Mexico (Crone and Tolstoy 2010; Reddy et al. 2012). This oil spill has a unique coincidence that a portion of the oil was entrapped in deeper layers of ocean known as a deep-water plume of oil (Hazen et al. 2010; Camilli et al. 2010).

Metagenomic studies of the surface water at the contaminated site showed microbial diversity primarily composed of *Halomonas, Alteromonas*, *Pseudoalteromonas* and *Cycloclasticus*, (Redmond and Valentine 2012; Gutierrez et al. 2013). Whereas the deep-water microbial diversity primarily composed of cryophilic oil-degrading microorganisms such as *Oceanospirillales, Cycloclasticus* and *Colwellia* (Hazen et al. 2010; Redmond and Valentine 2012). The deep-water plume of oil comprised a large population of *Oceanospirillales*, i.e. more than 90% of microbial diversity (Mason et al. 2012). Shotgun metagenomic sequencing study revealed expression of various set of genes involved in chemotaxis and biodegradation of hydrocarbons in deep water plume than compared to clean deep-water control (Mason et al. 2012). In addition to these genes, some genes responsible for BTEX (benzene, toluene, ethylbenzene and xylene) compound metabolism were also expressed at low levels. The dominant role of genes in *Oceanospirillales* sp. in the degradation of alkanes and cycloalkanes was indicated at early time points in the oil spill with the usage of single-cell genomics.

1.4.7 Increase Uranium Oxidation in the Contaminated Aquifer Through Biostimulation

Anthropogenic activities such as uranium mining, refining and processing as well as from natural environment are a significant contributor to uranium contamination in groundwater (Brugge and Buchner 2011). Mobilization of uranium from contaminated water by the microbes is achieved by altering its redox state (Newsome et al. 2014). The metal-reducing microbes reduce the aqueous soluble form of Uranium ($_{92}U^{238}$) to insoluble form, i.e. from U(VI) to U(IV) by supplying electron-donating compounds such as hydrogen, lactate, acetate and ethanol (Lovley et al. 1991). The best example of biostimulation studies was executed at US DOE Rifle place in Colorado, where processed uranium from mining site leached and contaminated the groundwater (Anderson et al. 2003). Stimulation with acetate, ethanol and emulsified vegetable oil was tested. The dominance of *Geobacter* sp. was found in the polluted groundwater (Anderson et al. 2003; Chandler et al. 2010; Chang et al. 2005). Stable isotope probing with ^{13}C-labelled acetate combined with 16S rRNA sequencing indicated the dominant role of *Geobacter* sp. in utilizing and incorporating the acetate into the biomass (Kerkhof et al. 2011). The whole-genome microarray study showed expression of *rpsC* gene in *Geobacter uraniireducens* to regulate the growth of bacterium and metabolism of uranium (Holmes et al. 2013). Biostimulation through amendment with emulsified vegetable oil (EVO) showed high numbers of *Geobacter* sp. in the total microbial profile along with high numbers of *Desulforegula* and *Pelosinus* spp. The *Pelosinus* strain UFO1 found at the Oak Ridge location was able to reduce soluble form of Uranium (${}_{92}U^{238}$) [U(VI)] to insoluble form [U(IV)] (Brown et al. 2014; Ray et al. 2011). GeoChip analysis during EVO amendment and after amendment indicated successive shifts in functional potential of the microbial population (Zhang et al. 2015). These shifts comprised of expression of genes for EVO degradation, reduction of U (VI), Fe (III), Mn (IV), SO_4^- and NO_3^-.

1.5 Conclusion

With the advancement of sequencing technology, the costs have declined, and dependence of researchers have boosted to the new level. The usage of high-throughput 16S rRNA sequencing in metagenomics is a vital tool for studying the diverse microbes in environmental samples. The best part of the metagenomic approach is the fact that with minimal invasion, location-specific, rapid remediation strategies can be developed. The study of application-oriented genes through metagenomics has a vital role in delineating the enzymatic pathways. Apart from these advancements, some of the critical information still needs exploration. This includes the information about the key taxa and elements of the metabolic pathways involved in the bioremediation process, mechanistic understanding of the community response, potential site-specific microbial biosensors and efficient prediction models with high accuracy.

References

Acinas SG, Antón J, Rodríguez-Valera F (1999) Diversity of free-living and attached bacteria in offshore western Mediterranean waters as depicted by analysis of genes encoding 16S rRNA. Appl Environ Microbiol 65:514–522

Ackert L (2012) Sergei vinogradskii and the cycle of life: from the thermodynamics of life to ecological microbiology. Springer, New York, pp 1850–1950

Alexander R, Kagi RI, Singh RK, Sosrowidjojo IB (1994) The effect of maturity on the relative abundances of cadalene and isocadalene in sediments from the Gippsland Basin, Australia. Org Geochem 21:115–120

Anderson RT, Vrionis HA, Ortiz-Bernad I et al (2003) Stimulating the in situ activity of *Geobacter* species to remove uranium from the groundwater of a uranium-contaminated aquifer. Appl Environ Microbiol 69:5884–5891

Antizar-Ladislao B (2010) Bioremediation: working with bacteria. Elements 6:389–394

Aubel D, Renaud FNR, Freney J (1997) Genomic diversity of several *Corynebacterium* species identified by amplification of the 16S–23S rRNA gene spacer regions. Int J Syst Evol Microbiol 47:767–772

Begon M, Harper JL, Townsend CR (1986) Ecology. Individuals, populations and communities. Blackwell Scientific Publications, Boston

Bending GD, Lincoln SD, Sørensen SR et al (2003) In-field spatial variability in the degradation of the phenyl-urea herbicide isoproturon is the result of interactions between degradative *Sphingomonas* spp. and soil pH. Appl Environ Microbiol 69:827–834

Beveridge TJ (2001) Use of the gram stain in microbiology. Biotech Histochem 76:111–118

Błaszczyk D, Bednarek I, Machnik G et al (2011) Amplified ribosomal DNA restriction analysis (ARDRA) as a screening method for normal and bulking activated sludge sample differentiation. Polish J Environ Stud 20:29–36

Blevins SM, Bronze MS (2010) Robert Koch and the 'golden age' of bacteriology. Int J Infect Dis 14:e744–e751

Borneman J, Triplett EW (1997) Molecular microbial diversity in soils from eastern Amazonia: evidence for unusual microorganisms and microbial population shifts associated with deforestation. Appl Environ Microbiol 63:2647–2653

Bowditch BM, Albright DG, Williams JGK, Braun MJ (1993) Use of randomly amplified polymorphic DNA markers in comparative genome studies. In: Methods in enzymology. Elsevier, New York, pp 294–309

Brown SD, Utturkar SM, Magnuson TS et al (2014) Complete genome sequence of *Pelosinus* sp. strain UFO1 assembled using single-molecule real-time DNA sequencing technology. Genome Announc 2:e00881

Brugge D, Buchner V (2011) Health effects of uranium: new research findings. Rev Environ Health 26:231–249

Camilli R, Reddy CM, Yoerger DR et al (2010) Tracking hydrocarbon plume transport and biodegradation at Deepwater Horizon. Science 330:201–204

Cardenas F, Alvarez E, de Castro-Alvarez M-S et al (2001) Screening and catalytic activity in organic synthesis of novel fungal and yeast lipases. J Mol Catal B Enzym 14:111–123

Chandler DP, Kukhtin A, Mokhiber R et al (2010) Monitoring microbial community structure and dynamics during in situ U (VI) bioremediation with a field-portable microarray analysis system. Environ Sci Technol 44:5516–5522

Chang Y-J, Long PE, Geyer R et al (2005) Microbial incorporation of 13C-labeled acetate at the field scale: detection of microbes responsible for reduction of U (VI). Environ Sci Technol 39:9039–9048

Christakopoulos P, Katapodis P, Kalogeris E, et al (2003) Antimicrobial activity of acidic xylo-oligosaccharides produced by family 10 and 11 endoxylanases. Int J Biol Macromol 31:171–175

Colwell RR, Brayton P, Herrington D et al (1996) Viable but non-culturable *Vibrio cholerae* O1 revert to a cultivable state in the human intestine. World J Microbiol Biotechnol 12:28–31

Cowan D, Meyer Q, Stafford W et al (2005) Metagenomic gene discovery: past, present and future. Trends Biotechnol 23:321–329

Crone TJ, Tolstoy M (2010) Magnitude of the 2010 Gulf of Mexico oil leak. Science 330:634

Cui C, Shu W, Li P (2016) Fluorescence in situ hybridization: cell-based genetic diagnostic and research applications. Front Cell Dev Biol 4:89

Cunliffe M, Kertesz MA (2006) Effect of *Sphingobium yanoikuyae B1* inoculation on bacterial community dynamics and polycyclic aromatic hydrocarbon degradation in aged and freshly PAH-contaminated soils. Environ Pollut 144:228–237

Czarny J, Staninska-Pięta J, Piotrowska-Cyplik A et al (2019) Assessment of soil potential to natural attenuation and autochthonous bioaugmentation using microarray and functional predictions from metagenome profiling. Ann Microbiol 69:945–955

da Silva Araújo SC, Silva-Portela RCB, de Lima DC et al (2020) MBSP1: a biosurfactant protein derived from a metagenomic library with activity in oil degradation. Sci Rep 10:1–13

Darwin C (2012) On the origin of the species and the voyage of the beagle. Graphic Arts Books, Portland

Dejonghe W, Berteloot E, Goris J et al (2003) Synergistic degradation of linuron by a bacterial consortium and isolation of a single linuron-degrading *Variovorax* strain. Appl Environ Microbiol 69:1532–1541

Desai C, Pathak H, Madamwar D (2010) Advances in molecular and "-omics" technologies to gauge microbial communities and bioremediation at xenobiotic/anthropogen contaminated sites. Bioresour Technol 101:1558–1569

Dubinsky EA, Conrad ME, Chakraborty R et al (2013) Succession of hydrocarbon-degrading bacteria in the aftermath of the deepwater horizon oil spill in the Gulf of Mexico. Environ Sci Technol 47:10860–10867

Ellilä S, Bromann P, Nyyssönen M et al (2019) Cloning of novel bacterial xylanases from lignocellulose-enriched compost metagenomic libraries. AMB Express 9:124

Escobar-Zepeda A, Vera-Ponce de Leon A, Sanchez-Flores A (2015) The road to metagenomics: from microbiology to DNA sequencing technologies and bioinformatics. Front Genet 6:348

Falkowski PG, Barber RT, Smetacek V (1998) Biogeochemical controls and feedbacks on ocean primary production. Science 281:200–206

Fischer SG, Lerman LS (1979) Length-independent separation of DNA restriction fragments in two-dimensional gel electrophoresis. Cell 16:191–200

Fischer SG, Lerman LS (1983) DNA fragments differing by single base-pair substitutions are separated in denaturing gradient gels: correspondence with melting theory. Proc Natl Acad Sci 80:1579–1583

Fisher MM, Triplett EW (1999) Automated approach for ribosomal intergenic spacer analysis of microbial diversity and its application to freshwater bacterial communities. Appl Environ Microbiol 65:4630–4636

Gich FB, Amer E, Figueras JB et al (2000) Assessment of microbial community structure changes by amplified ribosomal DNA restriction analysis (ARDRA). Int Microbiol 3:103–106

Gillespie DE, Brady SF, Bettermann AD et al (2002) Isolation of antibiotics turbomycin A and B from a metagenomic library of soil microbial DNA. Appl Environ Microbiol 68:4301–4306

Green SJ, Prakash O, Jasrotia P et al (2012) Denitrifying bacteria from the genus *Rhodanobacter* dominate bacterial communities in the highly contaminated subsurface of a nuclear legacy waste site. Appl Environ Microbiol 78:1039–1047

Gutierrez T, Singleton DR, Berry D et al (2013) Hydrocarbon-degrading bacteria enriched by the deepwater horizon oil spill identified by cultivation and DNA-SIP. ISME J 7:2091–2104

Handelsman J (2004) Metagenomics: application of genomics to uncultured microorganisms. Microbiol Mol Biol Rev 68:669–685

Hazen TC, Dubinsky EA, DeSantis TZ et al (2010) Deep-sea oil plume enriches indigenous oil-degrading bacteria. Science 330:204–208

Henkel M, Müller MM, Kügler JH et al (2012) Rhamnolipids as biosurfactants from renewable resources: concepts for next-generation rhamnolipid production. Process Biochem 47:1207–1219

Henne A, Schmitz RA, Bömeke M et al (2000) Screening of environmental DNA libraries for the presence of genes conferring lipolytic activity on *Escherichia coli*. Appl Environ Microbiol 66:3113–3116

Holmes DE, Giloteaux L, Barlett M et al (2013) Molecular analysis of the in situ growth rates of subsurface *Geobacter* species. Appl Environ Microbiol 79:1646–1653

Horak CE, Snyder M (2002) ChIP-chip: a genomic approach for identifying transcription factor binding sites. In: Methods in enzymology. Elsevier, New York, pp 469–483

Iyer VR, Horak CE, Scafe CS et al (2001) Genomic binding sites of the yeast cell-cycle transcription factors SBF and MBF. Nature 409:533–538

Jaouadi B, Abdelmalek B, Jaouadi Zaraî BNS (2011) The bioengineering and industrial applications of bacterial alkaline proteases: the case of SAPB and KERAB. InTech, London, pp 445–466

Jensen MA, Webster JA, Straus N (1993) Rapid identification of bacteria on the basis of polymerase chain reaction-amplified ribosomal DNA spacer polymorphisms. Appl Environ Microbiol 59:945–952

Kashyap R, Subudhi E (2014) A novel thermoalkaliphilic xylanase from Gordonia sp. is salt, solvent and surfactant tolerant. J Basic Microbiol 54:1342–1349

Katapodis P, Vardakou M, Kalogeris E, et al (2003) Enzymic production of a feruloylated oligosaccharide with antioxidant activity from wheat flour arabinoxylan. Eur J Nutr 42:55–60

Kavitha S, Selvakumar R, Sathishkumar M et al (2009) Nitrate removal using *Brevundimonas diminuta* MTCC 8486 from ground water. Water Sci Technol 60:517–524

Kerkhof LJ, Williams KH, Long PE, McGuinness LR (2011) Phase preference by active, acetate-utilizing bacteria at the Rifle, CO integrated field research challenge site. Environ Sci Technol 45:1250–1256

King GM, Kostka JE, Hazen TC, Sobecky PA (2015) Microbial responses to the Deepwater Horizon oil spill: from coastal wetlands to the deep sea. Annu Rev Mar Sci 7:377–401

Kirk JL, Beaudette LA, Hart M et al (2004) Methods of studying soil microbial diversity. J Microbiol Methods 58:169–188

Knapik K, Becerra M, González-Siso M-I (2019) Microbial diversity analysis and screening for novel xylanase enzymes from the sediment of the Lobios Hot Spring in Spain. Sci Rep 9:1–12

Ko K-C, Lee JH, Han Y, et al (2013) A novel multifunctional cellulolytic enzyme screened from metagenomic resources representing ruminal bacteria. Biochem Biophys Res Commun 441:567–572

Kumar BK, Balakrishnan H, Rele M V (2004) Compatibility of alkaline xylanases from an alkaliphilic Bacillus NCL (87-6-10) with commercial detergents and proteases. J Ind Microbiol Biotechnol 31:83–87

Kulkarni N, Shendye A, Rao M (1999) Molecular and biotechnological aspects of xylanases. FEMS Microbiol Rev 23:411–456

Li Y-H, Ding M, Wang J et al (2006) A novel thermoacidophilic endoglucanase, Ba-EGA, from a new cellulose-degrading bacterium, *Bacillus* sp. AC-1. Appl Microbiol Biotechnol 70:430–436

Li Y, Liu N, Yang H et al (2014) Cloning and characterization of a new β-Glucosidase from a metagenomic library of Rumen of cattle feeding with *Miscanthus sinensis*. BMC Biotechnol 14:85

Lim HK, Chung EJ, Kim J-C et al (2005) Characterization of a forest soil metagenome clone that confers indirubin and indigo production on *Escherichia coli*. Appl Environ Microbiol 71:7768–7777

Liu W-T, Marsh TL, Cheng H, Forney LJ (1997) Characterization of microbial diversity by determining terminal restriction fragment length polymorphisms of genes encoding 16S rRNA. Appl Environ Microbiol 63:4516–4522

Liu X, Ashforth E, Ren B et al (2010) Bioprospecting microbial natural product libraries from the marine environment for drug discovery. J Antibiot 63:415–422

Lorenz P, Liebeton K, Niehaus F, Eck J (2002) Screening for novel enzymes for biocatalytic processes: accessing the metagenome as a resource of novel functional sequence space. Curr Opin Biotechnol 13:572–577

Lovley DR, Phillips EJP, Gorby YA, Landa ER (1991) Microbial reduction of uranium. Nature 350:413–416

MacNaughton SJ, Stephen JR, Venosa AD et al (1999) Microbial population changes during bioremediation of an experimental oil spill. Appl Environ Microbiol 65:3566–3574

Maes N, De Gheldre Y, De Ryck R et al (1997) Rapid and accurate identification of Staphylococcus species by tRNA intergenic spacer length polymorphism analysis. J Clin Microbiol 35:2477–2481

Marcos MS, Lozada M, Dionisi HM (2009) Aromatic hydrocarbon degradation genes from chronically polluted Subantarctic marine sediments. Lett Appl Microbiol 49:602–608

Mason OU, Hazen TC, Borglin S et al (2012) Metagenome, metatranscriptome and single-cell sequencing reveal microbial response to deepwater horizon oil spill. ISME J 6:1715–1727

Maszenan AM, Liu Y, Ng WJ (2011) Bioremediation of wastewaters with recalcitrant organic compounds and metals by aerobic granules. Biotechnol Adv 29:111–123
McFall-Ngai M (2008) Are biologists in 'future shock'? Symbiosis integrates biology across domains. Nat Rev Microbiol 6:789–792
Megharaj M, Ramakrishnan B, Venkateswarlu K et al (2011) Bioremediation approaches for organic pollutants: a critical perspective. Environ Int 37:1362–1375
Miller JK, Barnes WM (1986) Colony probing as an alternative to standard sequencing as a means of direct analysis of chromosomal DNA to determine the spectrum of single-base changes in regions of known sequence. Proc Natl Acad Sci 83:1026–1030
Miyasaka T, Asami H, Watanabe K (2006) Impacts of bioremediation schemes on bacterial population in naphthalene-contaminated marine sediments. Biodegradation 17:227–235
Muyzer G, Smalla K (1998) Application of denaturing gradient gel electrophoresis (DGGE) and temperature gradient gel electrophoresis (TGGE) in microbial ecology. Antonie Van Leeuwenhoek 73:127–141
Myers RM, Maniatis T, Lerman LS (1987) Detection and localization of single base changes by denaturing gradient gel electrophoresis. In: Methods in enzymology. Elsevier, London, pp 501–527
Navarro E, Simonet P, Normand P, Bardin R (1992) Characterization of natural populations of *Nitrobacter* spp. using PCR/RFLP analysis of the ribosomal intergenic spacer. Arch Microbiol 157:107–115
Newsome L, Morris K, Lloyd JR (2014) The biogeochemistry and bioremediation of uranium and other priority radionuclides. Chem Geol 363:164–184
Philip B, D'Huys P-J, De Mot R, Springael D (2007) Characterization of novel linuron-mineralizing bacterial consortia enriched from long-term linuron-treated agricultural soils. FEMS Microbiol Ecol 62:374–385
Prosser JI (2015) Dispersing misconceptions and identifying opportunities for the use of 'omics' in soil microbial ecology. Nat Rev Microbiol 13:439–446
Qian C, Liu N, Yan X et al (2015) Engineering a high-performance, metagenomic-derived novel xylanase with improved soluble protein yield and thermostability. Enzym Microb Technol 70:35–41
Rastogi G, Sani RK (2011) Molecular techniques to assess microbial community structure, function, and dynamics in the environment. In: Microbes and microbial technology. Springer, Cham, pp 29–57
Ray AE, Bargar JR, Sivaswamy V et al (2011) Evidence for multiple modes of uranium immobilization by an anaerobic bacterium. Geochim Cosmochim Acta 75:2684–2695
Reddy CM, Arey JS, Seewald JS et al (2012) Composition and fate of gas and oil released to the water column during the deepwater horizon oil spill. Proc Natl Acad Sci 109:20229–20234
Redmond MC, Valentine DL (2012) Natural gas and temperature structured a microbial community response to the deepwater horizon oil spill. Proc Natl Acad Sci 109:20292–20297
Rees HC, Grant S, Jones B et al (2003) Detecting cellulase and esterase enzyme activities encoded by novel genes present in environmental DNA libraries. Extremophiles 7:415–421
Rhee J-K, Ahn D-G, Kim Y-G, Oh J-W (2005) New thermophilic and thermostable esterase with sequence similarity to the hormone-sensitive lipase family, cloned from a metagenomic library. Appl Environ Microbiol 71:817–825
Robleto EA, Borneman J, Triplett EW (1998) Effects of bacterial antibiotic production on rhizosphere microbial communities from a culture-independent perspective. Appl Environ Microbiol 64:5020–5022
Rosenbaum V, Riesner D (1987) Temperature-gradient gel electrophoresis: thermodynamic analysis of nucleic acids and proteins in purified form and in cellular extracts. Biophys Chem 26:235–246
Saxena G, Chandra R, Bharagava RN (2016) Environmental pollution, toxicity profile and treatment approaches for tannery wastewater and its chemical pollutants. In: Reviews of environmental contamination and toxicology. Volume 240, Springer, pp 31–69

Scheinert P, Krausse R, Ullmann U et al (1996) Molecular differentiation of bacteria by PCR amplification of the 16S–23S rRNA spacer. J Microbiol Methods 26:103–117
Sharma PK, Chand D (2012) Pseudomonas sp. xylanase for clarification of Mausambi and Orange fruit juice. Int J Adv Res Technol 1:1–3
Shi S, Bending GD (2007) Changes to the structure of *Sphingomonas* spp. communities associated with biodegradation of the herbicide isoproturon in soil. FEMS Microbiol Lett 269:110–116
Silva CC, Hayden H, Sawbridge T, et al (2012) Phylogenetic and functional diversity of metagenomic libraries of phenol degrading sludge from petroleum refinery wastewater treatment system. AMB express 2:1–13
Singh BK, Walker A, Morgan JAW, Wright DJ (2003) Effects of soil pH on the biodegradation of chlorpyrifos and isolation of a chlorpyrifos-degrading bacterium. Appl Environ Microbiol 69:5198–5206
Singh R, Chopra C, Gupta VK et al (2015) Purification and characterization of CHpro1, a thermotolerant, alkali-stable and oxidation-resisting protease of Chumathang hotspring. Sci Bull 60:1252–1260
Sleator RD, Shortall C, Hill C (2008) Metagenomics. Lett Appl Microbiol 47:361–366
Soni R, Chadha B, Saini HS (2008) Novel sources of fungal cellulases of thermophilic/thermotolerant for efficient deinking of composite paper waste. Bioresources 3:234–246
Spiegelman D, Whissell G, Greer CW (2005) A survey of the methods for the characterization of microbial consortia and communities. Can J Microbiol 51:355–386
Staley JT, Konopka A (1985) Measurement of in situ activities of non-photosynthetic microorganisms in aquatic and terrestrial habitats. Annu Rev Microbiol 39:321–346
Techtmann SM, Hazen TC (2016) Metagenomic applications in environmental monitoring and bioremediation. J Ind Microbiol Biotechnol 43:1345–1354
Tiedje JM (1995) Approaches to the comprehensive evaluation of prokaryote diversity of a habitat. Microb Divers Ecosyst Funct CAB Int Oxon, United Kingdom 73–87
Tiquia SM (2010) Using terminal restriction fragment length polymorphism (T-RFLP) analysis to assess microbial community structure in compost systems. In: Bioremediation. Springer, New York, pp 89–102
Turnbaugh PJ, Gordon JI (2009) The core gut microbiome, energy balance and obesity. J Physiol 587:4153–4158
Wang F, Li F, Chen G, Liu W (2009) Isolation and characterization of novel cellulase genes from uncultured microorganisms in different environmental niches. Microbiol Res 164:650–657
Watson DB, Kostka JE, Fields MW, Jardine PM (2004) The Oak Ridge field research center conceptual model. NABIR F Res Center, Oak Ridge
Wong KK, Tan LU, Saddler JN (1988) Multiplicity of beta-1, 4-xylanase in microorganisms: functions and applications. Microbiol Rev 52:305
Woo PCY, Lau SKP, Teng JLL et al (2008) Then and now: use of 16S rDNA gene sequencing for bacterial identification and discovery of novel bacteria in clinical microbiology laboratories. Clin Microbiol Infect 14:908–934
Yeh Y-F, Chang SC, Kuo H-W et al (2013) A metagenomic approach for the identification and cloning of an endoglucanase from rice straw compost. Gene 519:360–366
Zhang P, Wu W-M, Van Nostrand JD et al (2015) Dynamic succession of groundwater functional microbial communities in response to emulsified vegetable oil amendment during sustained in situ U (VI) reduction. Appl Environ Microbiol 81:4164–4172

The Coming Together of Sciences: Metagenomics for Microbial Biochemistry

2

Jyotsana Sharma, Sarmeela Sharma, Indu Sharma, Chirag Chopra, and Varun Sharma

Abstract

Microbes are essential for the smooth functioning of life as every life process in the biosphere involves the contribution of microbes. Previously, the study of the microbes has been primarily centred over laboratory-based pure-culture techniques due to which the extensive understanding regarding the microbial communities lags far behind. Metagenomics presents a novel strategy for examining the microbial populations that has upgraded the whole scenario of microbiological research. It involves the application of genomic analysis to whole microbial communities thereby clearly avoiding the necessity of isolation and culturing of individual members of the microbial community. It constitutes techniques like high-throughput sequencing, shotgun metagenomic sequencing of amplicon-based assays, gene prediction, metatranscriptomics and statistical studies, thus, making it feasible for the researchers to investigate the functional as well as metabolic diversity of microbiome by combining the power of genomics, bioinformatics and systems biology. It is a novel technique that can accommodate the analysis of genomes of many organisms concurrently.

J. Sharma · S. Sharma
School of Biotechnology, Shri Mata Vaishno Devi University, Kakryal, Jammu and Kashmir, India

I. Sharma · V. Sharma (✉)
Ancient DNA Laboratory, Birbal Sahni Institute of Palaeosciences, Lucknow, Uttar Pradesh, India

C. Chopra
School of Bioengineering and Biosciences, Lovely Professional University, Phagwara, Punjab, India
e-mail: chirag.18298@lpu.co.in

R. S. Chopra et al. (eds.), *Metagenomics: Techniques, Applications, Challenges and Opportunities*, https://doi.org/10.1007/978-981-15-6529-8_2

Keywords

Microbes · Metagenomics · Function-based metagenomics · System-based metagenomics

2.1 Introduction

The field of microbiology has gone through a considerable transformation during the last few years that has resulted in a complete alteration of the outlook of microbiologists towards microorganisms and the ways to study them. Microbes are an essential part of human life (Stark 2010). Most of the vital processes that occur in the biosphere involve the contribution of the microbes and hence act as the backbone of every ecological system by controlling biogeochemical cycling of numerous elements that are essential for life on the planet. The biogeochemical cycles that convert the key elements of life, i.e. carbon, nitrogen, oxygen, phosphorous and sulphur into the biologically accessible forms are mainly dependent on the microbes (Vieites et al. 2008). Microbial communities are also involved significantly in providing essential nutrients to their hosts that includes plants as well as animals. They perform very vital functions that are needed to maintain life on earth, for example, extraction of energy from the food that we eat, remediation of the toxins that are produced naturally or due to the activities of the humans in the environment and also adds value to the food through fermentation (Young 2017). Earlier, the study of microbes was based on laboratory-based culturing methods that were focused mainly on single species in pure culture. Therefore, most of the information regarding microbes are mainly laboratory-based attained in the conditions of growing them in the synthetic media in pure culture without considering the ecological factor. The understanding that most microorganisms cannot be grown in pure culture compelled the microbiologists to search for other alternative strategies (Riesenfeld et al. 2004). With the development of genomics and molecular biology, this physiological knowledge has got a strong support of its fundamental genetic basis. Moreover, with the emergence of metagenomics, it now become possible to investigate microbial population in their own natural environments which has proven to be a great boon for the research in microbiology as well as in medicine by opening new channels of research through unique analyses of genome heterogeneity and thus providing more access to the microbial diversity and its thorough understanding (Riesenfeld et al. 2004; Sleator et al. 2008).

The term metagenomics is defined as the analysis of a collection of similar but not alike items. Environmental genomics and community genomics are other synonyms of metagenomics (Handelsman 2004). Understanding about metagenomes has been considerably improved by the use of next-generation sequencing and also advances in the construction of clone libraries as well as bioinformatics (Méndez-García et al. 2018). The first experiment of cloning in the phage vector has been reported in the year 1991 (Schmidt et al. 1991). After that, the construction of a metagenomic library with DNA obtained from an assortment of organisms from the laboratory enrichment cultures has also been reported. Clones expressing cellulolytic activity were found in these libraries, which were referred to as gene libraries, also known as zoolibraries (Healy et al. 1995). Another group of researchers reported the construction of libraries from prokaryotes from seawater where they have identified a 40-kb clone containing a 16S rRNA gene demonstrating that the clone was derived from an archaeon that had never been cultured (Stein et al. 1996). Likewise, many studies based on metagenomics were reported from time to time that have revolutionised the whole avenue of microbiological research. The chapter will cover different approaches of metagenomics analysis and will also throw some light on its contribution towards mankind by exploring the unexplored areas of microbiology.

2.2 Metagenomic Approaches for Analysis of Microbial Communities

Metagenomics is a comparatively recent introduction in science that has already generated much information about the uncultured population of microbes. The success of metagenomic approaches is attributed to the availability of high-throughput methods of DNA sequencing and the advanced computing capabilities that are required to analyse the millions of random sequences in the libraries. Metagenomics constitutes amplification, sequencing as well as the study of the hypervariable region of the 16S rRNA prokaryotic gene and other phylogenetic marker genes (Mathieu et al. 2013; Martín et al. 2014). The study by Metagenomics includes techniques like genomic DNA extraction, library construction, taxonomic composition analysis, shotgun sequencing, and statistical analysis. Metagenomics can be divided into two different approaches, namely sequence-based and function-based analysis of the uncultured microorganisms (Kennedy et al. 2011).

Function-based metagenomic approach assesses the biochemical as well as metabolic activities of interest by cloning the random DNA fragments in vectors in order to generate an expression library followed by the screening of the library with a specific substrate for a particular phenotype, e.g. salt tolerance or enzyme activity along with identification of the phylogenetic based origin of the cloned DNA (Courtois et al. 2003; MacNeil et al. 2001). It is a powerful approach to identify clones that express a specific function. Functional metagenomic analysis has recognized numerous novel antibiotics, genes responsible for antibiotic resistance, transporters (Na (Li)/H) as well as degradative enzymes (Healy et al. 1995; Majerník et al. 2001). The ultimate potential of this approach is that it does not rely on the genes of interest to be identified by the sequence analysis, thus has the power to classify completely new classes of genes for new functions. A significant limitation of this approach is the inefficient expression of some metagenomic genes in the host bacteria to be used for screening. For example, during the investigation of lipolytic clones derived from German soil reported that only one among the 730,000 clones displayed lipase activity. From the soil sample collected from North America, only 29 out of 25,000 clones from the DNA library expressed hemolytic activity (Henne et al. 2000). Thus, it can be concluded that there is a dearth of active clones, thus requires the progression of efficient screening and selections intended for finding new functions as well as bioactive molecules. Similarly, as bacterial genetics depends on selections in order to perceive events of low frequency, the approach of metagenomics will also be enhanced by looking for selectable phenotypes in order to expand the number of biologically active metagenomic clones that can be analyzed and further could also be used to put together a basic structure for function-based metagenomic analyses (Rondon et al. 2000).

Sequence-based metagenomics is based on sequencing of DNA demonstrating the whole environmental sample. With the application of Next Generation Sequencing, it becomes possible to acquire complex information and facts about the microbial organisms present in a sample. It involves large-scale screening of the clones for the highly conserved 16S rRNA genes for the objective of identification followed by sequencing of the entire clone with a gene of interest along with sequencing of the total metagenome on an extensive scale in order to explore phylogenetic anchors within the reconstructed genomes (Hoff et al. 2008). The strategy of Sequence-based metagenomics is highly preferred as it not focussed on a limited set of microorganisms. Instead, it presents extensive information regarding all potential activities of the microbial population represented by the metagenome. However, the addition of new data in genetic databases and advancement of various bioinformatics tools enhances the popularity of this strategy for the search of functional activities of interest. The primary function of this approach in the analysis of microbial population is the reconstruction of metabolic pathways and investigating their services in the ecosystem. For this, genome sequences of microorganisms of a specific community are determined followed by performing the assembly of contigs derived from individual metagenomic sequencing reads by using different algorithms as well as tools of bioinformatics like MetaVelvet and Meta-IDBA (Namiki et al. 2012; Peng et al. 2011). The compilation of phylogenetic markers is increasing, and thus with the rise in the markers diversity, the potential of Sequence-based metagenomics approach will also increase and consequently more fragments of unidentified DNA will be assigned to the particular organisms from which they are derived (Tyson et al. 2004; Venter et al. 2004).

2.3 Role of Metagenomics in Imparting Commercial Perspective to Microbial Biochemistry

The exploration of the microbial world by metagenomics has helped in disclosing the extent of genetic as well as biochemical diversity that exists in the biosphere (Chistoserdova 2009). This has indeed added to our knowledge about the types as well as the statistics of microbes that are accountable for the functioning of various important biological and geological cycles which in turn deeply enhance our understanding about climate change and vigour of the ecosystem. It can not only perform the identification of the species in each population; however, also endow with an insight into the microbial metabolic activities as well as their functional roles (Langille et al. 2013). The applications of metagenomics in the context of microbial biochemistry are given in Fig. 2.1. Besides, growing access to the microbial biodiversity through metagenomic also presents an abundance of potential applications in both biomedicine and industry. Metagenomics also revolutionized the pathogen

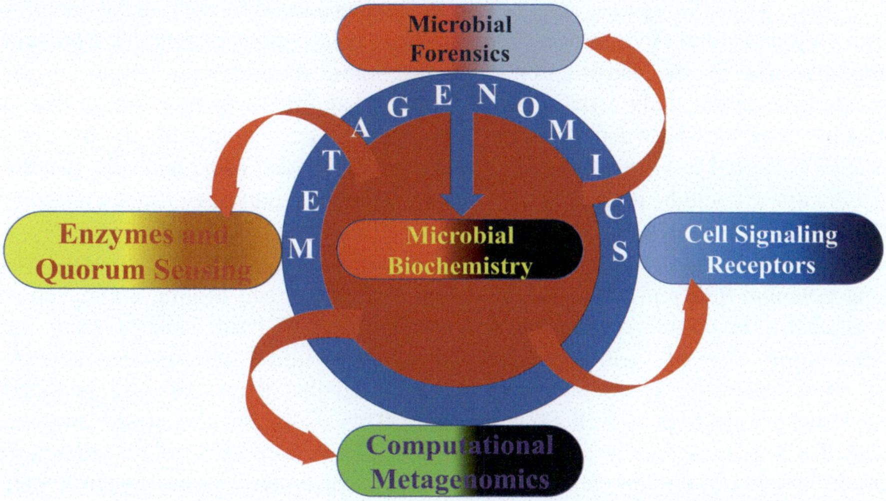

Fig. 2.1 Overview of the scope of metagenomics through microbial biochemistry

detection by allowing the concurrent detection of all microorganisms in a clinical sample by use of next-generation DNA sequencing. It has the potential to identify novel pathogens and find out their specific role in chronic human diseases. With recent studies, it has been shown that with the advancement in DNA sequencing and bioinformatics tools, it has become possible utilizing metagenomics to determine the whole genome sequences of the pathogens. This makes it possible to understand more thoroughly about antibiotic resistance and virulence (Miller et al. 2013). The analysis of the mechanism of action of the xenobiotics (particular the antibiotics) on the human gut microbiome has become very important in order to investigate the underlying mechanism of the drug resistance and also the gene associated with it to combat the problem of drug resistance and also the development of effective drugs that are less prone to pathogens.

Investigation of the mechanisms for resistance against xenobiotics and metabolism in the microbiome of the human gut will provide an insight of interaction between host and microbes, its biochemistry and an explanation to the variation observed in patient to patient response towards drugs. To some extent, this concern has been solved by metagenomics that has facilitated the analysis of microbial community for the aggregate genomes of the gut. Maurice et al. have reported how host-targeted drugs and antibiotics affected the gene expression as well as metabolic activity of a specific set of the active gut microbial population. These results pointed out the need of an assimilated characterization of the host, microbial and environmental factors that are involved in directing the response of the gut microbial community towards the xenobiotics; that could ultimately be used for the invent of various diagnostic tests or for developing therapeutic perspectives (Huttenhower et al. 2012; Maurice et al. 2013). Metagenomics is the most appropriate technology that has been contributing a lot in meeting the huge demand for novel enzymes and biocatalysts. Various industrially important enzymes (Cellulases, proteases, xylanases, lipases, amylases) have been produced through metagenomics from various natural environments like soil from cold regions or samples from any marine source by the construction of metagenomic libraries and further by the screening of biologically active clones (Lorenz et al. 2002). A novel lipase that is alkaline stable has been isolated by constructing a metagenomic library from the samples collected from marine sediments and has concluded with the results that the specific lipase could be used to impart a characteristic flavour as well as the odour in milk (Peng et al. 2014). Likewise, several papers have reported the isolation of lipases from different metagenomic libraries with novel characteristics and have great potential utility in the industrial sector (Hårdeman and Sjöling 2007; Selvin et al. 2012; Fu et al. 2013). Another alkaline pectate lyase isolated from the metagenome showed various properties like alkalophilic, high specific activity, and thermostability. These observations suggested that it can be used in many industrial prospective as well (Wang et al. 2014). Isolation of a novel amylase from the soil metagenome has been reported that showed 90% of its maximum activity even at low temperature (psychrophilic) which could prove their use very beneficial in various industrial applications (Sharma et al. 2010). It has been reported that screening of salt-tolerant microbial genes has been achieved by using the approach of functional metagenomics that could be used to produce various bioactive compounds in order to enhance the rate of crop production amid high saline conditions (Ahmed et al. 2018). These methodologies can provide new bits of knowledge into the ecologically important microbial networks and exercise that direct matter and energy transition on Earth. Such information in hand will be very beneficial in interpreting the relationship between biogeochemical cycles and human activities that collectively form the fortune of our planet. A detailed review of the enzymes discovered through metagenomics has been given in Chap. 7. Metagenomics has also provided insight into the various symbiotic associations as how microorganisms form symbiotic associations, communicate as well as acquire nutrition and produce energy with other organisms. Shotgun sequencing and reconstruction of metabolic pathway depicted that the symbionts are sulphur-oxidizing and sulphate-reducing bacteria which are proficient of carbon fixation and consequently capable of providing nutrition to the host organism (Woyke et al. 2006).

2.4 Quorum Sensing and Quorum Quenching Through Metagenomics

Bacteria communicate within their populations using quorum sensing (QS). The process of quorum sensing occurs through small signalling molecules called autoinducers (AI). Via these autoinducers, the microorganisms regulate luminescence, production of antibiotics, pathogenicity and growth patterns. The autoinducer molecules are of four types viz. acylhomoserine lactones (AHLs), alpha-hydroxyketones (AHKs), furanosylborate diesters (FBDE or AI-2) and autoinducer peptides (AIPs). One of the best studies mechanisms of quorum sensing mechanisms is the lux system of *Vibrio fischerii*. The QS system of *Vibrio fischerii* is regulated by the binding of the AHL to the intracellular receptor protein called luxR. luxR is a response regulator of transcription which is activated by binding of AHL. Activated luxR causes transcriptional activation of the luxI gene, which is responsible for production of the AHLs. This signalling works in a cell-density-dependent manner (Fuqua et al. 1994). The knowledge of the mechanism and gene products of the AHL-modulated QS systems is utilized for searching for the homologs of these genes in a microbial population. This is where the metagenomic approach finds significant applications. Hao and colleagues used the biosensor strain HC103 of Agrobacterium tumefaciens transformed with pJZ381 as a recipient strain for the metagenomic library. The metagenomic libraries was constructed from various soil and activated sludge samples (Wang et al. 2006). The metagenomic clones were transferred to the biosensor strain using conjugation method with the help of helper *E. coli* strain DH5a (pRK600). The Ti-plasmid of the sensor strain contained the traC-LacZ fusion, whereas the plasmid pJZ381 contained a pLac(lac promoter)-traR fusion. The screening principle was dependent on the presence of an autoinducer synthesizing gene in the metagenomic clones, which would lead to production of the AI. The AI would bind to the traR protein (response regulator). The activated traR would activate transcription of traC-LacZ fusion, causing production of blue coloured colonies on the agar plate. As a result, twenty-two blue colonies were identified from the screen that produced the quorum sensing molecules (Hao et al. 2010).

Another study on metagenomic approach was reported for the bioremediation application. Anammox (anaerobic ammonium oxidation) bioreactors are nitrogen recycling and removal systems that rely on conversion of ammonium ions to nitrites and nitrates without aeration. This leads to the removal of excess nitrogen without significant emission of greenhouse gases. Anammox bacteria comprise of genera like *Candidatus Anammoxoglobus*, *Candidatus Anammoximicrobium*, *Candidatus Brocadia*, *Candidatus Jettenia*, *Candidatus Kuenenia* and *Candidatus Scalindua* (Chu et al. 2015). These bacteria can grow symbiotically with ammonia and nitrite-oxidizing bacteria in anammox bioreactors. In the anammox bioreactors, unique biofilms are seen, which comprise of bacterial consortia. QS molecules, including N-octanoyl and N-hexanoyl homoserine lactones have been detected in such anammox systems (Tang et al. 2015). Biofilm formation in bacterial is a well-established concept and the mechanisms are also known (Dickschat 2010). Metagenomic approach was applied to the anammox bioreactor biofilms. The team extracted the metagenomic DNA from the biofilm samples and amplified the 16S rRNA genes. They sequenced the amplicons using the Illumina high-throughput sequencing platform and assembled the operational taxonomic units (OTUs). The phylogeny was studied using the EggNOG database. The sequence analysis revealed that apart from the anammox bacteria genera, other species of *Nitrospira*, and *Lautropia* were also detected. The details of these methods are explained in Chaps. 3 and 8.

Quorum quenching (QQ) or quorum-sensing inhibitors (QSIs) may include enzymes or other small molecules that inhibit the binding of AIs to the response regulators or cell-surface receptors (Dong and Zhang 2005). These inhibitors can also inhibit the process of biofilm formation in microbial communities.

Weiland-Bräuer and co-workers isolated nine QQ enzymes from the metagenome libraries (Weiland-Bräuer et al. 2016). These enzymes repressed the AHL and FBDE-mediated communication between *E. coli*, *Klebsiella oxytoca*, *P. aeruginosa*, *S. aureus* and *Bacillus subtilis*. The QQ enzymes also showed reduced biofilm formation in *Candida albicans and Staphylococcus epidermidis*. The QQ enzyme coded QQ7 also reduced the expression of transcriptional regulator icaR (Weiland-Bräuer et al. 2019).

Yaniv and colleagues used the BAC-metagenomic library constructed from the red sea samples (Sabehi et al. 2004) for screening of quorum sensing inhibitory molecules. The screening strategy was to check for violacein production by *Chromobacter violaceum*. The metagenomic clone with QSI activity could inhibit the production of violacein. Further screening for inhibition of biofilm formation was performed using the crude extracts from the positive clones identified from the violacein production assay. The team showed that the positive clones inhibited the biofilm formation in a dose-dependent manner (Yaniv et al. 2017). Thus, metagenomics likely holds the key to unlock the potential of microbial communities for providing beneficial QS molecules as well as QSIs.

2.5 Ion-Channels and Pumps Through Metagenomics

The ion-channels, pumps and sensors are important cell surface macromolecular machines, regulating the cell signalling and communication. Metagenomics has revealed the potential for the screening and identification of specific pump proteins called rhodopsins. The bacterial rhodopsins can function as light-driven H^+-pumps, sodium ion pumps and chloride ion pumps. Metagenomics provides access to these rhodopsins using the function as well as sequence-based approaches for screening. The screening strategy for a rhodopsin is a simple plate agar-based assay. The metagenomic clone that express a rhodopsin gene, would show orange-red pigmentation when cultured on agar-based media supplemented with retinal. The research carried out by Martinez and colleagues showed two prominent cloned expressing the light-operated proton pump (Martinez et al. 2007). More such studies have been carried out very recently. A unique group of bacterial rhodopsins was identified from a metagenomic screen in Israel. The team constructed and screened a fosmid-library to identify one positive clone using the all-trans-retinal screening (Pushkarev et al. 2018). The knowledge of these unique cell surface proteins can be obtained using the metagenomic approach.

2.6 Conclusion

Metagenomics has changed the approach of microbiologists towards the microbial research by redefining the whole concept of a genome and contributed immensely in the acceleration of the rate of gene discovery. Metagenomics has gained much success in analysing the vast microbiome of a given environmental sample. Various novel enzymes have been discovered through it that find ample applications in different areas. It has also led to the production of antibiotics and biosurfactants, which has given strength to the address of the issues like drug resistance, oil leakage, among others. Besides, issues of degradation of synthetic compounds which is very important for environmental protection from pollution has also been solved by the metagenomics. Therefore, it can be concluded that it provides a way to analyse the structural as well as functional genomics of the whole microbial diversity and hence plays an essential part in discovering novel genes for obtaining industrially valuable bioactive molecules and enzymes. Using the metagenomic approach, the quorum sensing molecules and the inhibitors of quorum sensing have been characterized. These applications enable the profiling of microbial communities in terms of their biochemical pathways as well as their physiology. Also, the discovery of bacterial rhodopsins through metagenomic approach can provide valuable insights into the biochemistry of microorganisms and microbial communities. Thus, we can say that metagenomics makes it possible to approach the massive diversity of the microbial population and has contributed a lot in understanding as well as gaining from the unculturable microbes.

References

Ahmed V, Verma MK, Gupta S et al (2018) Metagenomic profiling of soil microbes to mine salt stress tolerance genes. Front Microbiol 9:159

Chistoserdova L (2009) Functional metagenomics: recent advances and future challenges. Biotechnol Genet Eng Rev 26:335–352

Chu Z, Wang K, Li X et al (2015) Microbial characterization of aggregates within a one-stage nitritation–anammox system using high-throughput amplicon sequencing. Chem Eng J 262:41–48

Courtois S, Cappellano CM, Ball M et al (2003) Recombinant environmental libraries provide access to microbial diversity for drug discovery from natural products. Appl Environ Microbiol 69:49–55

Dickschat JS (2010) Quorum sensing and bacterial biofilms. Nat Prod Rep 27:343–369

Dong Y-H, Zhang L-H (2005) Quorum sensing and quorum-quenching enzymes. J Microbiol 43:101–109

Fu J, Leiros H-KS, de Pascale D et al (2013) Functional and structural studies of a novel cold-adapted esterase from an arctic intertidal metagenomic library. Appl Microbiol Biotechnol 97:3965–3978

Fuqua WC, Winans SC, Greenberg EP (1994) Quorum sensing in bacteria: the LuxR-LuxI family of cell density-responsive transcriptional regulators. J Bacteriol 176:269

Handelsman J (2004) Metagenomics: application of genomics to uncultured microorganisms. Microbiol Mol Biol Rev 68:669–685

Hao Y, Winans SC, Glick BR, Charles TC (2010) Identification and characterization of new LuxR/LuxI-type quorum sensing systems from metagenomic libraries. Environ Microbiol 12:105–117

Hårdeman F, Sjöling S (2007) Metagenomic approach for the isolation of a novel low-temperature-active lipase from uncultured bacteria of marine sediment. FEMS Microbiol Ecol 59:524–534

Healy FG, Ray RM, Aldrich HC et al (1995) Direct isolation of functional genes encoding cellulases from the microbial consortia in a thermophilic, anaerobic digester maintained on lignocellulose. Appl Microbiol Biotechnol 43:667–674

Henne A, Schmitz RA, Bömeke M et al (2000) Screening of environmental DNA libraries for the presence of genes conferring lipolytic activity on *Escherichia coli*. Appl Environ Microbiol 66:3113–3116

Hoff KJ, Tech M, Lingner T et al (2008) Gene prediction in metagenomic fragments: a large scale machine learning approach. BMC Bioinformatics 9:217

Huttenhower C, Gevers D, Knight R et al (2012) Structure, function and diversity of the healthy human microbiome. Nature 486:207

Kennedy J, O'leary ND, Kiran GS et al (2011) Functional metagenomic strategies for the discovery of novel enzymes and biosurfactants with biotechnological applications from marine ecosystems. J Appl Microbiol 111:787–799

Langille MGI, Zaneveld J, Caporaso JG et al (2013) Predictive functional profiling of microbial communities using 16S rRNA marker gene sequences. Nat Biotechnol 31:814

Lorenz P, Liebeton K, Niehaus F, Eck J (2002) Screening for novel enzymes for biocatalytic processes: accessing the metagenome as a resource of novel functional sequence space. Curr Opin Biotechnol 13:572–577

MacNeil IA, Tiong CL, Minor C et al (2001) Expression and isolation of antimicrobial small molecules from soil DNA libraries. J Mol Microbiol Biotechnol 3:301–308

Majerník A, Gottschalk G, Daniel R (2001) Screening of environmental DNA libraries for the presence of genes conferring Na+ (Li+)/H+ antiporter activity on Escherichia coli: characterization of the recovered genes and the corresponding gene products. J Bacteriol 183:6645–6653

Martín R, Miquel S, Langella P, Bermúdez-Humarán LG (2014) The role of metagenomics in understanding the human microbiome in health and disease. Virulence 5:413–423

Martinez A, Bradley AS, Waldbauer JR et al (2007) Proteorhodopsin photosystem gene expression enables photophosphorylation in a heterologous host. Proc Natl Acad Sci 104:5590–5595

Mathieu A, Delmont TO, Vogel TM et al (2013) Life on human surfaces: skin metagenomics. PLoS One 8:e65288

Maurice CF, Haiser HJ, Turnbaugh PJ (2013) Xenobiotics shape the physiology and gene expression of the active human gut microbiome. Cell 152:39–50

Méndez-García C, Bargiela R, Martínez-Martínez M, Ferrer M (2018) Metagenomic protocols and strategies. In: Metagenomics. Elsevier, New York, pp 15–54

Miller RR, Montoya V, Gardy JL et al (2013) Metagenomics for pathogen detection in public health. Genome Med 5:81

Namiki T, Hachiya T, Tanaka H, Sakakibara Y (2012) MetaVelvet: an extension of Velvet assembler to de novo metagenome assembly from short sequence reads. Nucleic Acids Res 40:e155–e155

Peng Y, Leung HCM, Yiu S-M, Chin FYL (2011) Meta-IDBA: a de Novo assembler for metagenomic data. Bioinformatics 27:i94–i101

Peng Q, Wang X, Shang M et al (2014) Isolation of a novel alkaline-stable lipase from a metagenomic library and its specific application for milkfat flavor production. Microb Cell Factories 13:1

Pushkarev A, Inoue K, Larom S et al (2018) A distinct abundant group of microbial rhodopsins discovered using functional metagenomics. Nature 558:595–599

Riesenfeld CS, Schloss PD, Handelsman J (2004) Metagenomics: genomic analysis of microbial communities. Annu Rev Genet 38:525–552

Rondon MR, August PR, Bettermann AD et al (2000) Cloning the soil metagenome: a strategy for accessing the genetic and functional diversity of uncultured microorganisms. Appl Environ Microbiol 66:2541–2547

Sabehi G, Béjà O, Suzuki MT et al (2004) Different SAR86 subgroups harbour divergent proteorhodopsins. Environ Microbiol 6:903–910

Schmidt TM, DeLong EF, Pace NR (1991) Analysis of a marine picoplankton community by 16S rRNA gene cloning and sequencing. J Bacteriol 173:4371–4378

Selvin J, Kennedy J, Lejon DPH et al (2012) Isolation identification and biochemical characterization of a novel halo-tolerant lipase from the metagenome of the marine sponge *Haliclona simulans*. Microb Cell Factories 11:72

Sharma S, Khan FG, Qazi GN (2010) Molecular cloning and characterization of amylase from soil metagenomic library derived from Northwestern Himalayas. Appl Microbiol Biotechnol 86:1821–1828

Sleator RD, Shortall C, Hill C (2008) Metagenomics. Lett Appl Microbiol 47:361–366

Stark LA (2010) Beneficial microorganisms: countering microbephobia. CBE—Life Sci Educ 9:387–389

Stein JL, Marsh TL, Wu KY et al (1996) Characterization of uncultivated prokaryotes: isolation and analysis of a 40-kilobase-pair genome fragment from a planktonic marine archaeon. J Bacteriol 178:591–599

Tang X, Liu S, Zhang Z, Zhuang G (2015) Identification of the release and effects of AHLs in anammox culture for bacteria communication. Chem Eng J 273:184–191

Tyson GW, Chapman J, Hugenholtz P et al (2004) Community structure and metabolism through reconstruction of microbial genomes from the environment. Nature 428:37–43

Venter JC, Remington K, Heidelberg JF et al (2004) Environmental genome shotgun sequencing of the Sargasso Sea. Science 304:66–74

Vieites JM, Guazzaroni M-E, Beloqui A et al (2008) Metagenomics approaches in systems microbiology. FEMS Microbiol Rev 33:236–255

Wang C, Meek DJ, Panchal P et al (2006) Isolation of poly-3-hydroxybutyrate metabolism genes from complex microbial communities by phenotypic complementation of bacterial mutants. Appl Environ Microbiol 72:384–391

Wang H, Li X, Ma Y, Song J (2014) Characterization and high-level expression of a metagenome-derived alkaline pectate lyase in recombinant *Escherichia coli*. Process Biochem 49:69–76

Weiland-Bräuer N, Kisch MJ, Pinnow N et al (2016) Highly effective inhibition of biofilm formation by the first metagenome-derived AI-2 quenching enzyme. Front Microbiol 7:1098

Weiland-Bräuer N, Malek I, Schmitz RA (2019) Metagenomic quorum quenching enzymes affect biofilm formation of *Candida albicans* and *Staphylococcus epidermidis*. PLoS One 14: e0211366

Woyke T, Teeling H, Ivanova NN et al (2006) Symbiosis insights through metagenomic analysis of a microbial consortium. Nature 443:950–955

Yaniv K, Golberg K, Kramarsky-Winter E et al (2017) Functional marine metagenomic screening for anti-quorum sensing and anti-biofilm activity. Biofouling 33:1–13

Young VB (2017) The role of the microbiome in human health and disease: an introduction for clinicians. BMJ 356:831

Part II

Applications of Metagenomics

Metagenomic DNA Sequencing: Technological Advances and Applications

3

Daljeet Singh Dhanjal, Chirag Chopra, and Reena Singh Chopra

Abstract

Over time, sequencing determination of the unknown isolate has become the routine process to gather information about any piece of the nucleic acid. There have been significant innovations in the sequencing technique, from sequencing one nucleotide at a time to the massively parallel sequencing methods. Over the past three decades, the Sanger sequencing method has been the sole sequencing approach. From the past 15 years, there has been a significant reduction in the DNA sequencing cost which has increased the demand to a revolutionary extent as the newer techniques are providing the inexpensive, robust and precise genomic information. The conventional and next-generation techniques have tremendously progressed and created wealthy opportunities in different research areas, especially in metagenomic approaches, as sequencing play a significant role in its analysis. Therefore, this chapter intends to provide information about past applied and current DNA sequencing approaches and will discuss their limitations and strength. Moreover, the application of these sequencing techniques will also be delineated.

Keywords

Heliscope · Illumina · Ion-torrent · Metagenomics · Next-generation sequencing · Oxford nanopore

D. S. Dhanjal · C. Chopra · R. S. Chopra (✉)
School of Bioengineering and Biosciences, Lovely Professional University, Phagwara, Punjab, India
e-mail: chirag.18298@lpu.co.in; reena.19408@lpu.co.in

R. S. Chopra et al. (eds.), *Metagenomics: Techniques, Applications, Challenges and Opportunities*, https://doi.org/10.1007/978-981-15-6529-8_3

3.1 Introduction

Ever since the introduction of microbes by Leeuwenhoek to their characterization via molecular approaches has changed the definition of microbial communities (Hugerth and Andersson 2017), various eminent scientists have researched isolating "invisible" microbes by conventional techniques. For more than 300 years, conventional techniques were the only option to investigate and gain information regarding the microbial world (Mora et al. 2016). However, in the 1970s, Carl Woese revolutionized the way of characterizing and identifying the microbes by integrating the ribosomal RNA genes and Sanger Sequencing method, which open the gates for the exploitation of uncultured microbial communities (Walters and Knight 2014). This further led to the development of gene expression techniques, which opened the new avenues for discovering new genes and metabolic products of industrial importance (Moody 2001). These achievements have set the base of a new field named "metagenomics" which theoretically means a collection of whole genomes from the specific environment without culturing it in in vitro conditions. In metagenomics, sequencing is an utmost important step in the process (Dhanjal and Sharma 2018).

Therefore, focusing on the sequencing, it has been found that for approximately two decades "Sanger Sequencing" approach has remained dominant and helped us in achieving monumental achievements including HGP (human genome project) (Chan 2005). Despite the various advances during this time, the drawbacks of automated Sanger sequencing inspired to update and improve the sequencing techniques for sequencing the genomes. The progress has made sequence readout robust, reliable and cheaper (Pareek et al. 2011; Heather and Chain 2016).

This chapter intends to provide the information regarding Sanger sequencing method and latest ('third- and next-generation') sequencing techniques, which have substituted Sanger-based method for numerous applications. The focus of this chapter is to understand the principle of the next-generation sequencing (NGS) platforms like Illumina, Ion Torrent, HeliScope, Pacific Biosciences (PacBio), 454/Roche, Sequencing by Oligo Ligation Detection (SOLiD) and Oxford Nanopore (ON). Furthermore, we will also discuss the application of NGS in biological and medical science in brief.

3.2 The Importance of DNA Sequencing

The importance of DNA sequencing approaches is that it provides valuable information about the variety of sources of biological importance (Kahvejian et al. 2008). Most common application of this approach is to reveal the patterns of genetic variation of the microbial community. Moreover, it also allows to gather other biological information like the function of particular genes based on homology, allow us to compare the gene to assess the microbial diversity and identify the metabolic pathway (Kunin et al. 2008). Most importantly, this approach enables us to get better insight about all organisms up to genomic level, i.e. evolution and potential of organisms on decoding the DNA sequence (Frazer et al. 2003). Now,

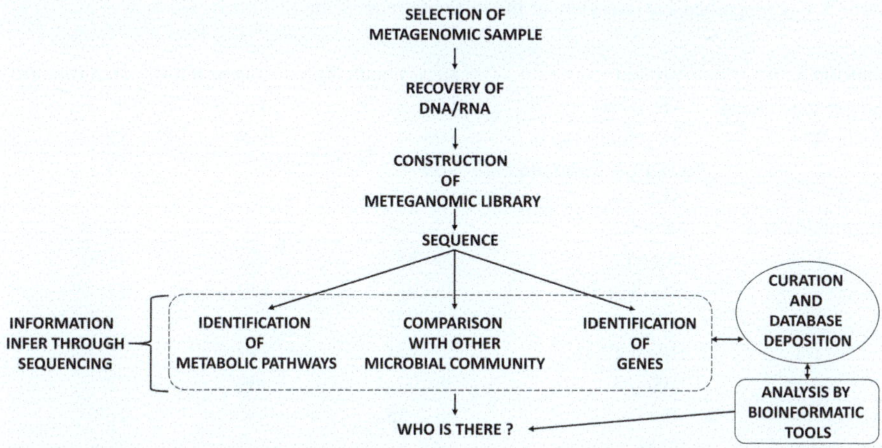

Fig. 3.1 The work of metagenomic approach highlighting the significance of sequencing

because of sequencing, we have genome sequences of about five hundred archaea as well as bacteria, fungi, nematodes, cow, chimpanzee, dog, animal models (fruit flies and zebrafish), plant model (*Arabidopsis thaliana*) and many other eukaryotes (National Research Council (US) Committee on Metagenomics: Challenges and Applications F 2007). As the metagenomic approach relies on gathering adequate sequence information in order to characterize the complete genome (Boolchandani et al. 2019). Major metagenomic projects involve the assembling of a complete genome from environmental sample (a mixture of samples), which further demands the physical recovery of organism-specific clones from environmental-DNA (eDNA) sequence databases comprising overlapping target-organism-specific contigs/sequences or eDNA libraries (Oniciuc et al. 2018). Indeed, the metagenomic field is now gaining considerable attention for gathering the ecological and evolutionary information about microbial ecosystem, mostly relies on the sequencing technique (Handelsman 2004). The brief outlook of the metagenomic approach has been depicted in Fig. 3.1. Therefore, improvement in sequencing technologies will aid in advancing our understanding of different biological pathways and change the available tools for metagenomic analysis (Zhang et al. 2011).

3.3 Different Sequencing Approaches: Evolution and Current State

Decoding the information stored in DNA molecules, which is the linear nucleotide base sequence, began in 1977 with the publishing of two major paper. One paper was published by Maxam and Gilbert, stating about the chemical degradation approach for DNA sequencing. Whereas Sanger and his colleagues published another paper, in which they presented the method founded on DNA synthesis and highlighted the use of dideoxy-nucleotides (Heather and Chain 2016). After these first-generation

Table 3.1 Comprehensive overview of the different sequencing platforms (Knetsch et al. 2019)

Sequencing platform	Version	Read length (bp)	Output (Gb)	Run time (h)
Sanger sequencing	ABI 3730	650	0	2
Roche 454	GS junior system	400	0.04	10
	GS FLX Titanium XL+	700	0.7	23
Illumina/solexa	MiniSeq	75–300	7.5	7–24
	MiSeq	50–600	15	4–56
	NextSeq 500	75–300	120	11–29
	HiSeq 2500	50–250	500	40–264
	HiSeq 4000	50–300	750	24–84
	HiSeq X	300	900	72
SoLiD	v4	50 + 35	1.4	168
Ion torrent	PGM	400	2.2	4–7
	Proton I	200	16	4
HeliScope	HeliScope	200	15	240
PacBio	RS II	12,000	0.66	6
	Sequel	10,000	3.85	6
Oxford nanopore	MinION	10,000	0.66	48

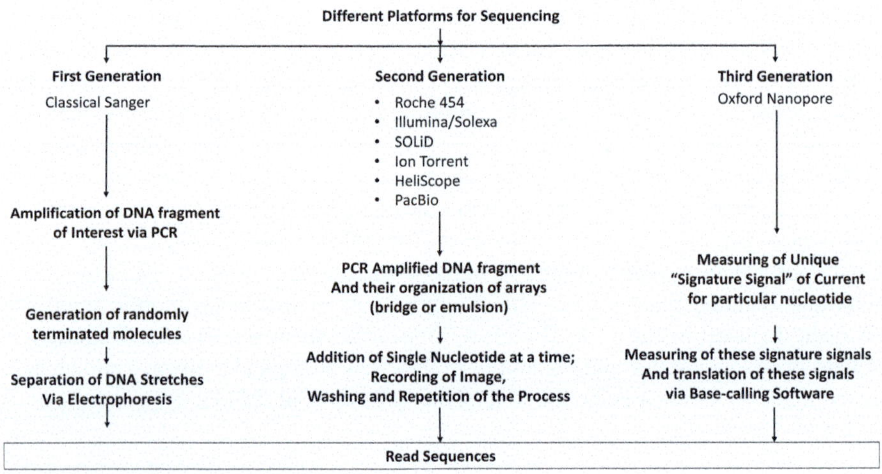

Fig. 3.2 General overview of different generations of sequencing platforms

sequencing techniques, numerous next-generation sequencing approaches were developed. Table 3.1 comprehends information on different sequencing platforms. The major sequencing approaches have been illustrated in Fig. 3.2.

3.3.1 Sanger Sequencing

Sanger Sequencing is said to be a first-generation sequencing approach which was introduced in 1977 and is also termed as "Chain Termination Method" (Totomoch-Serra et al. 2017). This sequencing technique involves three basic principles: One is DNA polymerization (the subsequent adding of new nucleotides complementary to template strand). The primer used for the sequencing is generally complementary to the template strand and decides the region on the template DNA for sequencing, as it contains 3′-OH end. This 3′-OH end of the primer functions as the initial point for the synthesis of a new strand. The catalytic activity of DNA polymerase aids in synthesizing the new DNA strand, which requires the standard four deoxyribonucleotide triphosphates (dNTPs) (Shendure et al. 2017). Except these, it also contains modified nucleotides called fluorescent dideoxyribonucleotide triphosphates (ddNTPs) in it. In opposition to the normal dNTPs, these modified ddNTPs lack the 3′-OH group, which are prerequisite for the addition of an adjacent nucleotide.

Meanwhile, these ddNTPs being fluorescently labelled allows their detection by the detector (França et al. 2002). As DNA polymerase is unable to distinguish between dNTPs and ddNTPs. Therefore, these dNTPs and ddNTPs randomly gets incorporated on the synthesizing DNA, as per their available concentration within the reaction mixture. Considering that the concentration of ddNTPs is found to be in lower concentration in comparison to dNTPs. Therefore, the synthesizing DNA stretches get terminated at variable length (Metzker 2005).

Secondly, the DNA product obtained after the reaction is separated according to their size with the help of electrophoresis on agarose gel competent to separate DNA strand of varied length. At last, DNA runs on the gel are visualized (França et al. 2002). Previously, this was done by using the radiolabeled ddNTPs and exposing the X-ray film to the gel. However, currently, fluorescent dye labelled ddNTPs are used, which emit light of varied wavelength (as different nucleotide emits different colour) when excited with a laser. Then generated peaks by fluorescent nucleotides are noted on a chromatogram. The chromatogram contains information about the newly formed DNA, which can be comprehended. Hence, each newly synthesized DNA sequence can also be stated as "reads" (Soper et al. 2003).

3.3.2 Roche 454

Roche 454, another sequencing platform working on sequencing-by-synthesis approach, entered the market in 2005 (Kulski 2016). It is one of the pyrosequencing techniques, as it detects the pyrophosphate released during the incorporation of each nucleotide in the newly synthesized DNA strand. Hence, it said to be the second-generation sequencing approach (Ari and Arikan 2016). In this approach, the DNA templates remain adhered to the microbeads containing immobilized oligonucleotides and get amplified with the help of emulsion PCR. The microbeads are then transferred to picotiter plate (PTP), and sequencing is initiated, which involves the activation of a series of downstream reactions, generating optical signal on the incorporation of each nucleotide. The detection of this light emission signals on the incorporation of each nucleotide allows us to deduce of a sequence of DNA fragment (Gilles et al. 2011). The major benefit of using a picotiter plate is that it allows us to run thousands of reactions in parallel and significantly increases the overall sequencing throughput. The reads generated by this approach are relatively long, i.e. up to 800 bp, which eases the mapping with the reference genome (Moorthie et al. 2011). The error in the result of this sequencing appears due to the presence of the homopolymer region leading to the deletion, insertion and substitution errors in the read. There are different variants of Roche 454 Sequencing like GS-20, GS-FLX, GS-FLX Titanium, GS-FLX+. The latest variant GS-FLX+ available has shown significant improvement as it generates read of 1000 bp length and produce >1 million reads in one run (Luo et al. 2012).

3.3.3 Illumina/Solexa

In 2007, another second-generation sequencing approach entered the commercial market, which also works on the sequencing-by-synthesis approach. There are different variants like Genome Analyzer I and II, HiSeq, and MiSeq are available in the market (Ambardar et al. 2016). This sequencing protocol starts with the ligation of the template DNA with the adaptor sequence on both ends, and later they are hooked on a glass flow cell. In the next step, the DNA template is amplified by subjecting it to bridge amplification, where each template generates approximately 1000 copies of the template. With the help of 3' inactivated fluorescent nucleotides and isothermal polymerase, the solitary bases get incorporated on synthesized DNA in each cycle. Hence, on the addition of a solitary base in each cycle is excited by the laser step. The emitted light signals are detected with the help of coupled-charge device camera (CCDC), which transfers this data to computer programs. Then, these computer programs translate these signals into nucleotide sequence (Thermes 2014). This approach overcomes the limitations of Roche 454 induced by homopolymer sequences, and additionally, this approach has the error rate between 1 and 2%. The major drawback of this sequencing approach is that it requires high precision during sample loading as overloading lead to the development of overlapping clusters during sequencing, which accounts for poor sequencing quality. Additionally, nucleotide substitution is the most common error deduced during this approach (Reuter et al. 2015).

3.3.4 SOLiD

Another second-generation sequencing entered the market as an update to Roche 454. Life Technologies first marketed this sequencer and later was acquired by Applied Biosystems in 2007. This approach adopts the technology of two-based sequencing followed by ligation. Moreover, the library construction module uses the same technology that of the Roche 454 Technology (Berglund et al. 2011). In this

DNA is sheared to fragments, which are then attached to oligonucleotide adapters that are further attached on the beads and cloned by emulsion PCR. After the denaturation step, the template adhering beads are enriched and placed on a solid substrate. The 3′ end of the template selected for attachment on the bead is modified for covalent attachment to the solid substrate. After this sequencing reaction is initiated via hybridizing the oligonucleotide sequence of the primer, complementary to the adapter sequence at the junction site. As fluorescently labelled 8-mers are consecutively ligated to DNA fragments, they emit the fluorescence on excitation and signal is recorded. The output comes in the form of color space, signifying the particular nucleotide, as four fluorescent colors are employed to allude 16 combinations of two nucleotide bases (Hodkinson and Grice 2015). This sequencing approach works based on ligation chemistry. Hence, ligation is persistently repeated, and in every new cycle, there is the removal of a complementary strand. In the new cycle, the sequencing initiates at the n-1 position of the template. This process is repeated until each base gets sequence twice. Later, the data recorded as colors are translated into nucleotides to deduce the sequence of DNA fragment (Voelkerding et al. 2009).

This sequencing technique can generate the shortest read of 35 bp and output of 3 Gb (giga base-pairs) in one run, and continuous efforts are being made to increase the read length to 75 bp and output to 30 Gb in one run. The accuracy of this sequencing technique is high precise as each base is read twice. Whereas, the error occurs during with long run time and with short reads. Additionally, the error also occurs due to the generation of noise during ligation cycle leading to an error of base identification, predominantly substitution error (Zhao et al. 2017).

3.3.5 Ion Torrent

Ion Torrent introduced another second-generation sequencing approach by the name "Ion Personal Genome Machine (PGM)". PGM is still a sequencing during synthesis but works with a novel approach. This approach uses the semiconductor-sequencing approach, as it detects the variation in pH because of the release of the proton during the polymerase reaction of DNA sequencing (Srinivasan and Batra 2014). During the process of sequencing, PGM recognizes whether the addition of nucleotide took place or not. At every step, the chip was filled with one type of nucleotide subsequently and change in voltage was noted. If there is no incorporation of nucleotide, it shows no voltage change. Whereas, when two same nucleotide gets incorporated, it shows a doubled voltage reading. This is the first sequencing machine which does not need any camera and fluorescence scanning, hence reducing the cost, size of the machine and robust analysis. Presently, it provides a 200 bp stretch of DNA in 2 h. Additionally, sample preparation takes less than 6 h. Moreover, it allows running the eight samples in parallel (Merriman et al. 2012).

In this sequencing platform, templates of sequences are generated on beads through emulsion PCR (emPCR). The water-oil emulsion is generated to separate small reaction vesicles which ideally comprises all the reagents required for sequencing, one sphere, and one library molecule. In this, there are two primers having a complementary sequence to that of library adapters, in which one is bound to the beads, and other is present in the solution. This enables them to select the library molecules containing A and B adapter, whereas excluding those molecules having two A or B adapters during emPCR from stacking on the beads. Additionally, this confirms even orientation of the sequence library molecules on the beads. Through emPCR steps, individual library molecules amplify themselves to millions of duplicate copies bounded to bead for its detection (Marine et al. 2019).

This process shows the error between the range of 1–1.7%, largely containing indels linked with under- or over-called flows. Moreover, the error rate increases with flow cycle and found to be non-homogeneous throughout the cycle. Furthermore, this platform provides the information of high-frequency indels (HFIs), where given base in the reference, share a majority of the same indel with aligned reads signifying the genuine difference among the sample and reference genomes. HFIs appears after every 1–2 kb in contrast to the reference sequence (Yeo et al. 2012).

3.3.6 HeliScope

Another second-generation sequencing approach which has entered in the sequencing market is known by the name "HeliScope" in which sequencing of a single molecule of DNA takes place. This DNA sequencing approach was made commercialized in 2007 by the company "Helicos Biosciences" (Pareek et al. 2011). The working principle of this approach is based on the principle of "true single-molecule sequencing (tSMS)" approach. For this, there is need for the preparation of DNA library via DNA shearing and followed by poly-(A) tail addition on the generated sheared DNA fragments, which are further hybridized to poly-(T) oligonucleotides attached to the flow cell and are getting sequenced parallelly. The sequencing cycle involves the extension step with one fluorescently labelled nucleotide out of four, which is detected by HeliScope sequencer on its addition to the template sequence. Further, subsequent cleaving of chemical fluorophores initiates the next elongation cycle to start with left out fluorescently labelled nucleotide in order to determine the DNA sequence (Tripathi et al. 2016). This sequencing approach is capable of generating 28 Gb of sequencing data and can take around about eight days. It also used to synthesize short stretch of DNA of maximum 55 bases. Recently, Helicos Biosciences has developed "one base at a time" nucleotide technology which has allowed us to execute the homopolymer sequencing as well as to conduct the RNA sequencing (Thompson and Steinmann 2010).

3.3.7 Pacific Biosciences (PacBio)

Although, the second-generation sequencing approaches have shown major improvement over limitations of sanger sequencing approach like short read length, which makes it ineffective during data assembly as well as the determination of complex, methylated or iso-form region in the genome. The single-molecule real-time (SMRT) sequencing approach, developed by Pacific Biosciences (PacBio) has evolved as an effective solution to overcome the limitations of first- and second-generation sequencing. This sequencing method comes under the "Third-generation Sequencing" and is a real-time sequencing method in which there is no need for any pauses between reading steps. This step makes it distinct from other sequencing approaches (Rhoads and Au 2015).

In this sequencing approach, sequenced information is gathered while the replication process of the targeted DNA molecule takes place. It involves the single-stranded closed circular DNA, which acts as a template and is also known as "SMRTbell". The SMRTbell template is generated by ligating hairpin adaptors on both ends of a targeted double-stranded DNA (dsDNA). The working mechanism of this sequencing technique involves the loading of the SMRTbell sample on the chip, as known as SMRTcell. Later, SMRTbell diffuses through sequencing units, i.e. zero-mode waveguide (ZMW). Then, this ZMW imparts the smallest amount for detection. In every ZMW, a single polymerase is impaired at the bottom of the chip, which allows the binding of SMRTbell with template strand and initiates the replication process. The replication process taking place in all the ZMWs of SMRTcell are cataloged by measuring the light impulse, where each impulse of ZMW corresponds to the sequence of bases/continuous long read (CLR). As SMRTbell synthesis a closed circle, therefore on replication of one strand of target dsDNA, it incorporates the bases of adaptors before the other strand. Hence, the CLR generated can be split into multiple reads by determining and cutting out the adaptor sequence (Mccarthy 2010).

Further, the consensus sequence of multiple reads in one ZMW yield a circular consensus sequence (CCS) read with high precision. In the case of a large target DNA, the multiple reads/CCS are not generated instead of that single read output is obtained. Moreover, this approach allows us to analyze to study the base modification like methylation as kinetic variation in recorded in real-time (Gueidan et al. 2019).

3.3.8 ON

Oxford Nanopore is the third-generation sequencing approach which has entered in the sequencing market in 2014. It is the first portable miniature sequencing device named "MinION". This approach has separated itself from the philosophy of chain termination and polymerase-based sequencing, as it does not work on the principle of complementary-strand synthesis (Jain et al. 2016). Neither it records the fluorescence signals, therefore eliminating the need for optics and lasers. Hence, it is a small and cheap alternative device, as it doesn't require the need for a large number of chemicals and liquid handling. In its place, the sequencing takes place on the minuscule membrane. This approach is not new as it rooted to the idea in a distant past in 1989. The idea is that protein pores are united to a membrane, around which minute current is passed. As the ions, flow through these pores in the solution, causes the subsequent obstruction in current flow. However, when macromolecule blocks the pore temporarily, they increase the resistance within the system, which can be determined (Lu et al. 2016). Considering the difference in affinity, moving speed and shape of these macromolecules blocking the pore, theoretically, leaves the unique "signature" signal at the current level. Each nucleotide present in DNA produces a different signal. Hence, feeding of ssDNA molecule via suitable pore will generate the sequence of dips and spikes of current called Squiggles. These squiggles can further be translated to nucleotide sequence through base-calling software (Plesivkova et al. 2019).

Lately, various protein pores have been discovered which can be used in nanopore nucleotide sensors, and now even holes present in sheets of graphene can be used for DNA sequencing with this approach. Initially, Oxford Nanopore used the modified α-haemolysin pore synthesized by *Staphylococcus aureus*, which was then embedded onto the polymer membrane. With this, using the complementary tether oligonucleotides, single-stranded DNA (ssDNA) are concentrated nearby the membrane. Though the current passing through the membrane promotes the translocation of nucleic acid via pore. Moreover, DNA molecules are forced to move towards the pores by motor protein. Furthermore, in real-time the current levels are measured by integrated circuit as well as cloud-based base-caller, which aid in constructing sequence for analysis by bioinformatic tools (downstream) while the other molecules are still undergoing sequencing (Kono and Arakawa 2019).

The MinION Mk1 sequencer contains disposable flow cells having 2048 pores, out of which 512 pores can be employed for co-current sequencing. During the sequencing process, the nucleotide strings pass through the pore at 70 bp/s (basepairs per second), where these pores sustain for 48 h, generating 6 Gb run (theoretically). Though all the pores are not occupied at all times, therefore sequencing run of 1–2 Gb generates a good yield. But, read lengths synthesized are limited because of the DNA sample quality, as single nick in long double-stranded DNA (dsDNA) stretch leads to in reading truncation. Although this method is not fully matured and has a high error rate, continuous improvement in the approach is reducing this rate. Even the errors generated are not completely arbitrary. Hence, few intrinsic-biases in base-calling occurs, which are difficult to be corrected as MinION can run two sequencing passes simultaneously as per its design. However, continuous updates are being made to improve this platform and address the sequencing issues (Kono and Arakawa 2019). As, PromethION modular design containing 48 flow cells attached in parallel fashion, add each cell encompasses 3000 Nanopore channels. This allows the individual or concurrent running of flow cells, hence increasing the capacity and flexibility significantly (Leggett and Clark 2017). Whereas, GridION is another handheld-sized device for sequencing having five MinION flow cells connected in a parallel fashion. Moreover, this product has a dedicated computing module on-board, which allow it to complete on-device base-calling and prevent the requirement of extra IT infrastructure. Even though this approach is in its infancy but continuous are being made to improve the potential of portable real-time sequencing.

VolTRAX and SmidgION are the latest version developed by Oxford Nanopore, which are on the verge of their release (Wang et al. 2014).

3.4 Applications of Sequencing Techniques

Even though the sequencing of gene and genome are significant applications of sequencing techniques, but these approaches have a wide range of application and few applications have been comprehended and discussed below.

3.4.1 Role in Clinical Microbiology

The different sequencing approaches were introduced and claimed as the solution to diverse diagnostic problems associated with microbiology. In spite of the various applications like antimicrobial resistance profiling, genotyping, outbreak analysis and pathogen identification, the next-generation sequencing approaches remained less explored (Gu et al. 2019). Which was surprising, as the advances in this approach have made the sequencers faster and affordable. However, there are two challenges which obstruct the use of NGS in microbiology is the complex data and absence of robust technology to interpret the obtained data. As it is known that high-throughput whole-genome sequencing (HT-WGS) has attained the level and reliability to precisely describe the phylogenetic history and species population of emerging pathogens (Di Resta et al. 2018). Whereas, comparative analysis of WGS uncovers the variation in the genetic sequence and often define their role in the development of antimicrobial resistance or virulence. Furthermore, it enables us to characterize the strains based on single nucleotide, hence tremendously enhancing the precision power of this approach from a conventional approach like Amplified Fragment Length Polymorphism (AFLP), Denaturing or Temperature gradient gel electrophoresis (DGGE or TGGE), Pulsed-Field Gel Electrophoresis (PFGE), Restriction Fragment Length Polymorphism (RFLP) and Multiple-Locus Variable Number Tandem Repeat (ML-VNTR) Analysis (Fournier et al. 2014; Collineau et al. 2019). Intrinsically, WGS can be a major asset in managing pathogenic outbreak as well as infection prevention in the healthcare sector. Besides that this approach is more suitable for defining the outbreak coverage and trace the extent of MDR-pathogen at both regional and global level, which is an advancement over the conventional epidemiological investigation approaches (Koutsoumanis et al. 2019).

3.4.2 Role in Virology

Traditional characterization and detection of viruses via inoculation of cell culture and sample filtration are thought to be a time-consuming and labor-intensive approach. Although the approaches like cell-based assays cannot be wholly neglected, the robust detection by the molecular approach like quantitative reverse transcription PCR (RT-qPCR) has its own importance for clinical analysis (Leland and Ginocchio 2007). The onset of advancement in sequencing approaches has drastically changed the viral discovery by allowing detection along with the characterization of new viral strain without any former information on the primary sequence of DNA or RNA. Shotgun Metagenome Sequencing is one such approach in which DNA or RNA in a sample is sequenced. The available DNA or RNA are obtained from bacteria, viruses and sample matrix (body fluid) (Strazzulli et al. 2017). Another application is viral resistance characterization via amplicon deep sequencing, which precisely allows the detection and characterization of viral variants that could be associated with the development of resistance against direct-acting antivirals during therapy (Zhao et al. 2017). Nowadays, the NGS is predominantly used instead of the Sanger sequencing approach for detecting viral resistance. Also, data obtained via amplicon sequencing can be conducted in standardized manner and aid in submitting the data to regulatory bodies like EMA and FDA (Arias et al. 2018).

3.4.3 Role in Clinical Genetics

Sequencing of the Human genome has tremendous applications in clinical genetics. As genome sequencing means to gather the information regarding the arrangement of nucleotide in DNA stretch and the obtained sequencing further compared with reference human genome. The primary aim of which is to comprehend and find the difference in genetic make-up linked with the phenotype of diseased patient (Lander et al. 2001). Primarily, to detect the minute variations induced because of deletion/insertion/substitution of nucleotides, short-read NGS approach is used. Whereas, the analysis of significant structural variants relies on the length of the read as well as bioinformatics pipeline used (Tattini et al. 2015). Moreover, it is difficult to sequence the entire genome via amplicon sequencing in clinical analysis. Hence, targeted genes are sequenced, in which a probe is designed with oligonucleotides of targeted sequence and sequenced via NGS. This approach is also stated as "exome" sequencing (Kamps et al. 2017).

3.4.4 Role in Targeted Sequencing

As most of the research and diagnostics require the targeted gene or region of the genome for its assessment and does not require WGS as extra data will only be discarded. For instance, in case of diagnosis, there is a need for determining the presence of a short repeat of sequence at a particular location in genomic loci. In that case, the solution is simple, i.e. to amplify the sequence of interest through PCR and merely sequencing of that amplified fragment (Katsanis and Katsanis 2013). On the other hand, environmental and microbiology research, the advent of NGS has enabled the researchers to outline the composition of the whole microbial community or microbiome. The advent of NGS has enabled us to directly detect the multiple pathogens instantaneously (Fanning et al. 2017). The 16S rRNA ribotyping is now predominantly used as a genetic marker to know about bacteria present in diverse clinical material like feces, urine, gut, skin, soil, and sputum (Patel 2001). Moreover, degenerative primers are also used to amplify the conserved but mutable 16S rRNA-coding sequence. Hence on amplification, the amplified product obtained is sequenced. Later, the sequencing read obtained, characterized by comparing the reference database. In the end, after assessing the number of categorized reads of each taxonomy level, a relative description is generated, which provides information regarding the biodiversity of microbes in a sample (Clarridge 2004). Even microbiology and biochemical are there which provide information about the biodiversity, but they also have some limitations. The outmost limitation is the lack of variation in the targeted sequence. In most cases, because of the high similarity in sequence in variable regions of the 16S rRNA gene, the identification gets restricted to the genus level. Additionally, this 16S rRNA sequencing only provide the information regarding the bacteria and excludes the two important taxa i.e. fungi and viruses from evaluation (Janda and Abbott 2007).

All comprehended drawbacks can be overcome by applying the untargeted approach, in which the DNA will be directly isolated and sequenced from the sample also known as "Shotgun Metagenomics". Due to this, sequencing method is not any more challenging, but now it requires a combination of bioinformatic skill and computational resources, that are usually limited to dedicated sequencing laboratories (Alvarenga et al. 2017; Quince et al. 2017).

3.4.5 Role in Transcriptome Sequencing

As all the sequencing platforms are restricted for processing DNA and library preparation purpose of all the biomolecules which can be translated or associated with DNA. This also involves, DNA bound to a particular protein, methylated DNA and RNA (Zhang et al. 2018). An important application is the one in which reverse transcriptase is used to translate RNA to cDNA, which can be subsequently sequenced. Typically, all the type of RNA sequence like messenger RNA, micro RNA and non-coding RNA (Wang et al. 2009). In cases, where whole-genome sequencing of the particular species is too costly, in that case using the whole protein-coding transcriptome for determining the coding sequence is a brilliant alternative. On combing the NGS, cDNA sequencing is also stated as RNA seq and is frequently employed for determining the level of gene expression. For attaining this, a library of sequences should be prepared from reverse-transcribed copies of mRNA. The creates a library with a high number of cDNA transcripts in contrast to rare transcripts (Hrdlickova et al. 2017). In addition to that, ribosomal RNA is also the chief component of RNA, in a few cases, they account for 80–90% of total RNA. Hence, pre-depletion of this rRNA fraction is essential before sequencing in order to use the full capacity of sequencer for sequencing the informative stretch of RNA sample (Kim et al. 2019).

Because NGS can robustly process the different number of molecules in one go and outnumber a large number of genes, by merely counting the repetition of cDNA sequence and assess its gene expression level, typically, 10 million cDNA fragments are sequenced for 10–1000 transcripts. This is the reason why expression value determined by RNAseq is efficient than the old high-throughput methods like microarray and comparably precise and sensitive than quantitative PCR (Christodoulou et al. 2011).

3.5 Conclusion

Metagenomics has come up in a big way towards bioprospection of unique environmental niches for valuable small and macromolecules. As compared to the traditional microbiological approaches, the coverage of metagenomics is considerably higher, providing a greater scope for novel genes and ORFs. For unlocking such potential, the role of DNA sequencing is indispensable. With the advent of sophisticated sequencing techniques, high-throughput sequencing has become convenient and inexpensive. There are three generations of DNA sequencing techniques, which have been rigorously and routinely updated. From sanger's dideoxy sequencing to the minion (oxford nanopore), different chemistries have been utilized for sequencing the DNA with good quality reads and long read length. The various sequencing platforms have been utilized for sequencing the metagenomes for gene discovery and data analysis. Simply sequencing the DNA is not sufficient. Analysis and assembly of the sequence for meaningful analysis is equally critical. Sequence assembly tools are explained in detail in Chap. 13. The big-genome data is available in open-source databases and are usable for different applications in diverse fields of research. The application of high-throughput sequencing is well established in drug discovery, novel enzymes, microbial biodiversity, clinical microbiology, among others. These applications make high-throughput DNA sequencing platforms indispensable for biotechnological and biomedical research.

References

Alvarenga DO, Fiore MF, Varani AM (2017) A metagenomic approach to cyanobacterial genomics. Front Microbiol 8:809. https://doi.org/10.3389/fmicb.2017.00809

Ambardar S, Gupta R, Trakroo D et al (2016) High throughput sequencing: an overview of sequencing chemistry. Indian J Microbiol 56:394–404

Ari Ş, Arikan M (2016) Next-generation sequencing: advantages, disadvantages, and future. In: Plant omics: trends and applications. Springer, Cham, pp 109–135

Arias A, López P, Sánchez R et al (2018) Sanger and next generation sequencing approaches to evaluate HIV-1 virus in blood compartments. Int J Environ Res Public Health 15:1697. https://doi.org/10.3390/ijerph15081697

Berglund EC, Kiialainen A, Syvänen AC (2011) Next-generation sequencing technologies and applications for human genetic history and forensics. Investig Genet 2:23

Boolchandani M, D'Souza AW, Dantas G (2019) Sequencing-based methods and resources to study antimicrobial resistance. Nat Rev Genet 20:356–370

Chan EY (2005) Advances in sequencing technology. Mutat Res Fundam Mol Mech Mutagen 573:13–40

Christodoulou DC, Gorham JM, Kawana M et al (2011) Quantification of gene transcripts with deep sequencing analysis of gene expression (DSAGE) using 1 to 2 μg total RNA. Curr Protoc Mol Biol 25:9. https://doi.org/10.1002/0471142727.mb25b09s93

Clarridge JE (2004) Impact of 16S rRNA gene sequence analysis for identification of bacteria on clinical microbiology and infectious diseases. Clin Microbiol Rev 17:840–862

Collineau L, Boerlin P, Carson CA et al (2019) Integrating whole-genome sequencing data into quantitative risk assessment of foodborne antimicrobial resistance: a review of opportunities and challenges. Front Microbiol 10:1107

Dhanjal DS, Sharma D (2018) Microbial metagenomics for industrial and environmental bioprospecting: the unknown envoy. In: Microbial bioprospecting for sustainable development. Springer, Singapore, pp 327–352

Di Resta C, Galbiati S, Carrera P, Ferrari M (2018) Next-generation sequencing approach for the diagnosis of human diseases: open challenges and new opportunities. Electron J Int Fed Clin Chem Lab Med 29:4–14

Fanning S, Proos S, Jordan K, Srikumar S (2017) A review on the applications of next generation sequencing technologies as applied to food-related microbiome studies. Front Microbiol 8:1829

Fournier PE, Dubourg G, Raoult D (2014) Clinical detection and characterization of bacterial pathogens in the genomics era. Genome Med 6:1–15. https://doi.org/10.1186/s13073-014-0114-2

França LTC, Carrilho E, Kist TBL (2002) A review of DNA sequencing techniques. Q Rev Biophys 35:169–200

Frazer KA, Elnitski L, Church DM et al (2003) Cross-species sequence comparisons: a review of methods and available resources. Genome Res 13:1–12

Gilles A, Meglécz E, Pech N et al (2011) Accuracy and quality assessment of 454 GS-FLX Titanium pyrosequencing. BMC Genomics 12:245. https://doi.org/10.1186/1471-2164-12-245

Gu W, Miller S, Chiu CY (2019) Clinical metagenomic next-generation sequencing for pathogen detection. Annu Rev Pathol Mech Dis 14:319–338. https://doi.org/10.1146/annurev-pathmechdis-012418-012751

Gueidan C, Elix JA, McCarthy PM et al (2019) PacBio amplicon sequencing for metabarcoding of mixed DNA samples from lichen herbarium specimens. MycoKeys 53:73–91. https://doi.org/10.3897/mycokeys.53.34761

Handelsman J (2004) Metagenomics: application of genomics to uncultured microorganisms. Microbiol Mol Biol Rev 68:669–685. https://doi.org/10.1128/mmbr.68.4.669-685.2004

Heather JM, Chain B (2016) The sequence of sequencers: the history of sequencing DNA. Genomics 107:1–8

Hodkinson BP, Grice EA (2015) Next-generation sequencing: a review of technologies and tools for wound microbiome research. Adv Wound Care 4:50–58. https://doi.org/10.1089/wound.2014.0542

Hrdlickova R, Toloue M, Tian B (2017) RNA-Seq methods for transcriptome analysis. Wiley Interdiscip Rev RNA 8:1

Hugerth LW, Andersson AF (2017) Analysing microbial community composition through amplicon sequencing: from sampling to hypothesis testing. Front Microbiol 8:1561

Jain M, Olsen HE, Paten B, Akeson M (2016) The Oxford nanopore MinION: delivery of nanopore sequencing to the genomics community. Genome Biol 17:239. https://doi.org/10.1186/s13059-016-1103-0

Janda JM, Abbott SL (2007) 16S rRNA gene sequencing for bacterial identification in the diagnostic laboratory: pluses, perils, and pitfalls. J Clin Microbiol 45:2761–2764

Kahvejian A, Quackenbush J, Thompson JF (2008) What would you do if you could sequence everything? Nat Biotechnol 26:1125–1133

Kamps R, Brandão RD, van den Bosch BJ et al (2017) Next-generation sequencing in oncology: genetic diagnosis, risk prediction and cancer classification. Int J Mol Sci 18:308

Katsanis SH, Katsanis N (2013) Molecular genetic testing and the future of clinical genomics. Nat Rev Genet 14:415–426

Kim IV, Ross EJ, Dietrich S et al (2019) Efficient depletion of ribosomal RNA for RNA sequencing in planarians. BMC Genomics 20:1–12. https://doi.org/10.1186/s12864-019-6292-y

Knetsch CW, van der Veer EM, Henkel C, Taschner P (2019) DNA sequencing. In: Molecular diagnostics. Springer, Singapore, pp 339–360

Kono N, Arakawa K (2019) Nanopore sequencing: review of potential applications in functional genomics. Develop Growth Differ 61:316–326

Koutsoumanis K, Allende A, Alvarez-Ordóñez A et al (2019) Whole genome sequencing and metagenomics for outbreak investigation, source attribution and risk assessment of food-borne microorganisms. EFSA J 17:5898. https://doi.org/10.2903/j.efsa.2019.5898

Kulski JK (2016) Next-generation sequencing — an overview of the history, tools, and "omic" applications. In: Next generation sequencing - advances, applications and challenges. InTech, London

Kunin V, Copeland A, Lapidus A et al (2008) A bioinformatician's guide to metagenomics. Microbiol Mol Biol Rev 72:557–578. https://doi.org/10.1128/mmbr.00009-08

Lander ES, Linton LM, Birren B et al (2001) Initial sequencing and analysis of the human genome. Nature 409:860–921. https://doi.org/10.1038/35057062

Leggett RM, Clark MD (2017) A world of opportunities with nanopore sequencing. J Exp Bot 68:5419–5429. https://doi.org/10.1093/jxb/erx289

Leland DS, Ginocchio CC (2007) Role of cell culture for virus detection in the age of technology. Clin Microbiol Rev 20:49–78

Lu H, Giordano F, Ning Z (2016) Oxford nanopore MinION sequencing and genome assembly. Genomics Proteomics Bioinformatics 14:265–279

Luo C, Tsementzi D, Kyrpides N et al (2012) Direct comparisons of illumina vs. Roche 454 sequencing technologies on the same microbial community DNA sample. PLoS One 7: e30087. https://doi.org/10.1371/journal.pone.0030087

Marine RL, Magaña LC, Castro CJ et al (2019) Comparison of illumina MiSeq and the ion torrent PGM and S5 platforms for whole-genome sequencing of picornaviruses and caliciviruses. bioRxiv 705632. https://doi.org/10.1101/705632

Mccarthy A (2010) Third generation DNA sequencing: pacific biosciences' single molecule real time technology. Chem Biol 17:675–676

Merriman B, Torrent I, Rothberg JM (2012) Progress in ion torrent semiconductor chip based sequencing. Electrophoresis 33:3397–3417

Metzker ML (2005) Emerging technologies in DNA sequencing. Genome Res 15:1767–1776

Moody DE (2001) Genomics techniques: an overview of methods for the study of gene expression. J Anim Sci 79:E128. https://doi.org/10.2527/jas2001.79e-supple128x

Moorthie S, Mattocks CJ, Wright CF (2011) Review of massively parallel DNA sequencing technologies. HUGO J 5:1–12

Mora M, Mahnert A, Koskinen K et al (2016) Microorganisms in confined habitats: microbial monitoring and control of intensive care units, operating rooms, cleanrooms and the international space station. Front Microbiol 7:1573

National Research Council (US) Committee on Metagenomics: Challenges and Applications F (2007) From genomics to metagenomics: first steps. National Academies Press, Washington

Oniciuc EA, Likotrafiti E, Alvarez-Molina A et al (2018) The present and future of whole genome sequencing (WGS) and whole metagenome sequencing (WMS) for surveillance of antimicrobial resistant microorganisms and antimicrobial resistance genes across the food chain. Gene 9:268

Pareek CS, Smoczynski R, Tretyn A (2011) Sequencing technologies and genome sequencing. J Appl Genet 52:413–435

Patel J (2001) 16S rRNA gene sequencing for bacterial pathogen identification in the clinical laboratory. Mol Diagn 6:313–321. https://doi.org/10.1054/modi.2001.29158

Plesivkova D, Richards R, Harbison S (2019) A review of the potential of the MinION™ single-molecule sequencing system for forensic applications. Wiley Interdiscip Rev Forensic Sci 1: e1323. https://doi.org/10.1002/wfs2.1323

Quince C, Walker AW, Simpson JT et al (2017) Shotgun metagenomics, from sampling to analysis. Nat Biotechnol 35:833–844

Reuter JA, Spacek DV, Snyder MP (2015) High-throughput sequencing technologies. Mol Cell 58:586–597

Rhoads A, Au KF (2015) PacBio sequencing and its applications. Genomics Proteomics Bioinformatics 13:278–289

Shendure J, Balasubramanian S, Church GM et al (2017) DNA sequencing at 40: past, present and future. Nature 550:345–353. https://doi.org/10.1038/nature24286

Soper SA, Owens C, Lassiter S et al (2003) DNA sequencing using fluorescence detection. In: Topics in fluorescence spectroscopy. Kluwer Academic Publishers, Boston, pp 1–68

Srinivasan S, Batra J (2014) Four generations of sequencing- is it ready for the clinic yet. J Next Gener Seq Appl 1:1–8. https://doi.org/10.4172/2469-9853.1000107

Strazzulli A, Fusco S, Cobucci-Ponzano B et al (2017) Metagenomics of microbial and viral life in terrestrial geothermal environments. Rev Environ Sci Biotechnol 16:425–454. https://doi.org/10.1007/s11157-017-9435-0

Tattini L, D'Aurizio R, Magi A (2015) Detection of genomic structural variants from next-generation sequencing data. Front Bioeng Biotechnol 3:92

Thermes C (2014) Ten years of next-generation sequencing technology. Trends Genet 30:418–426

Thompson JF, Steinmann KE (2010) Single molecule sequencing with a HeliScope genetic analysis system. Curr Protoc Mol Biol 7:10

Totomoch-Serra A, Marquez MF, Cervantes-Barragán DE (2017) Sanger sequencing as a first-line approach for molecular diagnosis of Andersen-Tawil syndrome. F1000Research 6:11610. https://doi.org/10.12688/f1000research.11610.1

Tripathi R, Sharma P, Chakraborty P, Varadwaj PK (2016) Next-generation sequencing revolution through big data analytics. Front Life Sci 9:119–149. https://doi.org/10.1080/21553769.2016.1178180

Voelkerding KV, Dames SA, Durtschi JD (2009) Next-generation sequencing: from basic research to diagnostics. Clin Chem 55:641–658. https://doi.org/10.1373/clinchem.2008.112789

Walters WA, Knight R (2014) Technology and techniques for microbial ecology via DNA sequencing. In: Annals of the American Thoracic Society. American Thoracic Society, New York, p S16

Wang Z, Gerstein M, Snyder M (2009) RNA-Seq: a revolutionary tool for transcriptomics. Nat Rev Genet 10:57–63

Wang Y, Yang Q, Wang Z (2014) The evolution of nanopore sequencing. Front Genet 5:449. https://doi.org/10.3389/fgene.2014.00449

Yeo ZX, Chan M, Yap YS et al (2012) Improving indel detection specificity of the ion torrent PGM benchtop sequencer. PLoS One 7:e45798. https://doi.org/10.1371/journal.pone.0045798

Zhang J, Chiodini R, Badr A, Zhang G (2011) The impact of next-generation sequencing on genomics. J Genet Genomics 38:95–109

Zhang H, He L, Cai L (2018) Transcriptome sequencing: RNA-seq. In: Methods in molecular biology. Humana Press Inc, Totowa, pp 15–27

Zhao M, Liu D, Qu H (2017) Systematic review of next-generation sequencing simulators: computational tools, features and perspectives. Brief Funct Genomics 16:121–128. https://doi.org/10.1093/bfgp/elw012

Environmental Microbial Forensics: How Hidden is the Truth?

4

Peta Pavan Kumar, Kiran Yadav, Varun Vevek, Harshit Khandal, and Rajni Kumari

Abstract

The human society has witnessed bioterrorism and bio crime since biblical times. In the year 1155 A.D., Emperor Barbarossa poisoned water-wells with human bodies in Tortona, Italy. Since then, several instances of the use of pathogenic microbes to spread disease, have been witnessed to conquer nations or to practice terrorism. Easy availability of pathogens and involvement of low-cost methods have increased such instances. Managing risks and threats of bioterrorism and bio crime and strengthening biosecurity have become challenging tasks for governments and law-enforcement agencies. Microbial forensics is an arena of forensic science which applies the science of microbiology to forensic investigations. This field focuses on the characterization of evidence recovered from bioterrorism acts, bio crime, hoax or any inadvertent release of pathogens and biochemicals into the civilization. The forensic microbial investigation is like a routine forensic investigation as it also includes crime scene investigation, chain-of-custody, handling and bagging of evidence, analysis of evidence, interpretation of result and presentation in court. Different molecular biology-based assays such as nuclear acid amplification techniques, MLST (Multi Locus Sequence Typing), VNTR (Variable Number of Tandem Repeats) can be used to identify the microbial strains in the investigation process. This chapter focuses on the role of microbial forensics in enabling the law-enforcement bodies to deal with the menace of bioterrorism and bio crime systematically.

Keywords

Bio crime · Bioterrorism · Microbial forensic · MLST · Pathogen

P. P. Kumar · K. Yadav (✉) · V. Vevek · H. Khandal · R. Kumari
School of Bioengineering and Biosciences, Lovely Professional University, Phagwara, Punjab, India

R. S. Chopra et al. (eds.), *Metagenomics: Techniques, Applications, Challenges and Opportunities*, https://doi.org/10.1007/978-981-15-6529-8_4

4.1 Introduction

Investigating officers, victims and people who are known to the victims usually want to identify the person who committed the crime. Physical evidence recovered from the crime scene plays a vital role in the investigation. Handwriting identification, DNA analysis, fingerprint comparison, body-fluid identification, analysis of drugs and poisons have been well established in the field of forensic science (Budowle et al. 2005a). In the twenty-first century, the ability to manipulate and disperse pathogens increased which led to a profound concern regarding the potential use of microbial pathogens and their toxins for causing significant harm to the plants, humans and animals. Such types of threats can be identified and solved using forensic science techniques. One of the significant threats that can affect humankind is an outbreak of a severe disease either naturally or intentionally. Such type of event can create havoc in the society resulting in death or harm and also disruption to the economy. There are numerous bacteria, viruses and fungi that can cause serious health issues to humans, plants and animals. Using them as a potential source of bioweapon can disrupt the economic stability and political activities of a state or country. Easier access to technology has increased the likelihood of using bioweapons for creating mass destruction. In some journals, it has been cited that the use of bioweapon is more economical as compared to the traditional weapon (Murch 2003). Thus, bioterrorism poses a significant challenge in society. There are numerous examples of bioterrorism incidents in history using a pathogen to create disruption. Anthrax letter incident in 2001 established the need to strengthen homeland security and introduce forensic science approach for externalization and dissuasion (Budowle et al. 2005b).

Apart from bioterrorism, pathogens and toxins can be used to commit bio crime. Bio crimes are similar to traditional crimes in many ways, as the main focus is to harm an individual. In traditional crimes usually, gun, stone or knife is used while in bio crimes, a biological weapon is used to harm an individual or inflict an injury. Most of the pathogens that were used in the past to commit bio crimes were less lethal, for example, *Bacillus anthracis,* smallpox and Ebola, but they were easily accessible. During the fourteenth century, the European plague epidemic started as Tartan soldiers used dead bodies of their fellow soldiers suffering from plague to create havoc and invade in the walled city of Kaffa (Derbes 1966). It is impossible to anticipate the next biological agent that can be used to create damage or for illicit purpose.

Some harmful microbes are grown by providing the required environmental conditions while some are grown in the laboratory conditions. A virus can quickly spread from one person to another in a short period and cause infection to millions of people so that there is an economic loss to that country. Intentional use of microbes for killing people has been reported in many countries.

In microbial forensics, the epidemiological issues are to be identified and characterized for a specific disease caused by a pathogen or their toxins. Also, the forensic scientists analyze the mode of transmission of the pathogens and any changes that have been made intentionally to change their effect against the humans, animals and plant materials (Budowle et al. 2005b). The development of microbial forensics has been slower, relative to other disciplines of Forensic Sciences. Thus, it is considered as a minor program majorly in a variety of government agencies. Over the last few years, an acute awareness has developed regarding the threats of pathogen and toxin weapons which indicate microbial forensics has become necessary.

4.2 History of Pathogens

Biological agents have been used since the past centuries for spreading infection or for attacking the population of one's country. Anthrax caused by the bacterium *Bacillus anthracis* is used for biological crime due to its physical properties and virulence factors. Several countries have tried to use anthrax as a biological weapon in the military due to its potential (Pohanka and Kuča 2010). Anthrax is a rare bacterial infection caused by inhalation, ingestion, or if one comes in contact with the endospore of *Bacillus anthracis*. This bacterial species usually cause disease in herbivorous mammals and human beings who live near animals. Inhalational anthrax is a rare case, but a few cases related to inhalation anthrax were reported in the United States of America in 1978 (Aggarwal et al. 2011). In Russia, a military biological weapon facility in Sverdlovsk accidentally released the endospores of anthrax, and the case of anthrax appeared in humans that were even 4 km far away from the site. At the same time, some animals were also infected that were staying 50 km away from the site. By this way, we can say that pathogens can travel over long geographical distances. Globally there were two thousand cases due to anthrax, but only two cases were reported from the United States (Franz et al. 1997; Pohanka and Kuča 2010).

Plague is a contagious bacterial disease caused by *Yersinia pestis*. It is a zoonotic bacterium and usually found in small mammals and their fleas. There are three forms of plague infection in humans (Stenseth et al. 2008).

- If buboes are formed, then it is known as Bubonic plague,
- If the infection spreads to the lungs, then it is known as Pneumonic plague,
- If the infection spreads to bloodstream leading to systemic infection, then it is known as Septicemic form.

A common form of the plague is the pneumonic plague that killed over a million people across the globe. Many American army sailors were killed during the world war as they were moved to Boston from Philadelphia (Christopher et al. 1997). After 4 days, many people got infected and were hospitalized. More than one-quarter of Europe was infected. Usually, influenza kills the infants and elderly, but people who died due to this disease were mostly young men and women in their twenties and thirties. In other cases, the carcasses of animals and humans were used to contaminate the water supply of enemies. During the final months of the Second World War, Japan had planned to use plague as a biological weapon against United States citizens during Operation Cherry Blossom. Japan had planned to execute this plan on September 22, 1945, but it could not be executed due to the surrender of Japan on August 15, 1945 (Baumslag 2005). In 1972, Biological and Toxin Weapons Convention was signed by US, USSR, UK and several other nations to ban the development, production and stocking of microbes and their poisonous products except for amounts needed for peaceful research (Ligon 2006). This convention came into force on March 26, 1975 and is a multilateral treaty of infinite duration. One hundred sixty-five countries signed it. The significant terms of the treaty were to ban the stockpiling, acquisition, development and retention of toxins and biological agents of "types and quantities" that have no justification for protective and peaceful use. If any state after entering into the convention possesses any biological agent or toxin, then they have nine months to destroy it or divert it for peaceful use (Working 2001).

4.3 Pathogens in Microbial Forensics

4.3.1 Categorization of Pathogens in Microbial Forensics

Certain properties make microorganisms useful as a biological weapon. Some of these properties are as follows:

- easy accessibility
- culturability
- large-scale production capability
- stability during the production process
- toxicity
- virulence
- ability to retain potency after the production process (Greenwood 1997).

Based on the etiology of the disease and mortality, pathogens have been classified into three categories:

Category A: Pathogens that have been placed under this category show a high mortality rate and cause significant impacts on public health. In most of the cases, initially, a wild animal is infected and then the pathogen transfers to the human population. For example, *Clostridium botulinum* found in canned foods, *Yersinia pestis* found in rats are category A pathogens.

Category B: Pathogens that are placed in this group have less mortality rate than the category A pathogens. Examples: *Rickettsia prowazekii, Coxiella burnetii.*

Category C: Pathogens placed in this category have the potential for high mortality and morbidity rates and can cause significant health impacts. Example: Nipah virus, Hendra virus.

Some of the common viruses that have caused havoc in the health care sector are:

4.3.2 Nipah Virus

It spreads through bats and pigs. The people who are living near to the pigs are infected with the disease. It was first observed in Malaysia and Singapore (Paton et al. 1999). It can spread from one person to another and from animal to humans. It is a type of RNA virus from the genus Henipavirus. It is a newly emerged virus, so there is no treatment for this virus. It can be prevented by avoiding exposure to bats, infected pigs and by avoiding drinking raw date palm sap. Around 50% to 70% of the infected people died due to this virus since there was no cure. In India, it first spread in the state of Kerala. This virus has been named after a village in Malaysia, i.e. Sungai Nipah, where pigs were infected and later on many people got infected with this virus in the year 1999, to stop the spread of this virus, a massive number of pigs were euthanized (Syed 2018). In India and Bangladesh, the disease spread from one infected person to another. Symptoms usually appear after 5–14 days which can be fever, headache, drowsiness followed by mental illness.

4.3.3 Hanta Virus

ORTHOHANTAVIRUS or Hantavirus is a negative-sense RNA virus. It mostly infects rodents. When humans come in contact with infected rodent urine, saliva or faeces, they get infected. Hantavirus is named after the river Hantan that is present in South Korea.

4.3.4 Brucella

Brucella is the causative organism of brucellosis. It is a highly contagious zoonosis and is caused through unpasteurized milk or uncooked meat from infected animals. It is also known as Undulant fever, Malta fever and Mediterranean fever. It is a small, gram-negative, non-motile, non-spore forming rod-shaped bacteria. Since the twentieth-century Brucellosis has been recognized in humans and animals. Brucella species was used as a weapon in certain countries in the mid-twentieth century (Leitenberg 2001).

4.4 Role of Microorganisms in Forensic Investigations

Microbial forensics is a multidisciplinary field dedicated to analyzing evidence from bio crime or bioterrorism or unintentional toxin release for imputation purpose. The ultimate goal of imputation is to identify the person who was involved in bio crime. Apart from microbiological tools, traditional forensic tools and techniques such as DNA analysis, fingerprint analysis, and tool mark examination will also be used to investigate a bio crime. If we compare forensic microbiological evidence with other forensic evidence, then there is nothing exceptionally unique. Recognition of crime scene, preservation of crime scene, chain of custody, evidence collection, shipping of evidence, analysis of evidence, interpretation of result and presentation in court will be carried out in the same manner as it is done with other forensic evidence except that the evidence will be biohazardous. If there is no proper recognition of the crime scene, then it is quite impossible to identify the suspect or the nature of the incident, whether it was intentional or unintentional. Usually, bio crime and bioterrorism have been categorised in two parts, i.e. cases where crime has been committed in an undisguised manner or cases where crime has been committed covertly (Budowle et al. 2005a). Cases where no disguise is needed, one can place the pathogen in a public place or open space. In contrast, in covert cases unlikely occurrence of a particular disease will alarm the law enforcement agencies. In both cases, a partnership between public health and law enforcement is essential.

In order to plan better forensic investigation strategies, analysis of past cases can help in identifying critical problems and different approaches that should be adopted for active investigation. Some of the cases where epidemiological and molecular biology knowledge was used are Sverdlovsk anthrax case and the case of an HIV-infected dentist who might have infected patients with HIV (Jackson et al. 1998).

In the first case, sources of anthrax spores from an anthrax epidemic were explored. In April 1979 an outbreak of human anthrax occurred in Sverdlovsk, Russia. Although government officials attributed it to the consumption of contaminated meat, but western governments believed the cause to be a release of accidental spores from a nearby military research facility. At least two hundred patients died, and people affected from it lived within a narrow zone of approximately 4 km of south and east of military facility (Keim et al. 2011). Patient fatalities did not show any symptoms of skin anthrax; instead, they showed symptoms of pulmonary anthrax which ruled out the possibility of consumption of contaminated meat. Further investigation revealed that livestock from six different villages residing 50 km southeast of the facility also died due to anthrax. Tissue samples from 11 dead patients were examined. DNA extracted from tissue sample was analysed using the Polymerase Chain Reaction. B. anthracis toxin and capsular antigen genes required for pathogenicity were present in tissues from each of these victims.

In the second case, an attempt was made to determine the cause and source of HIV in patients that implicated a dentist based in Florida in 1980s (Keim et al. 2011). In the late 1980s, many people from Florida became HIV infected, although their lifestyle never put them in such kind of risk. Investigation indicated that HIV transmission occurred due to invasive dental care from a dentist with AIDS. The dentist was first identified as HIV positive in 1986. The actual mode of transmission of HIV was unknown, but the data implicated him as the source as the patients who got infected from HIV infection visited the same dentist. DNA sequences data from infected patient and dentist along with a local control group and outgroup were analysed. The analysis showed that HIV nucleotide sequences from several patients were closely related to the dentist.

4.4.1 Essential Components of Microbial Forensics Programs

- Proper identification and detection are essential for impeding bioterrorism. For effectively carrying out the attribution of an individual to a crime, robust analytical techniques should be developed, and proper implementation of those techniques needs to be done. Use of DNA based assays and analytical techniques of physics and chemistry can be used.
- Proper database and information will play an essential role in microbial forensics. Databases based on genomic sequences of bioagents and pathogenic agents need to be developed.
- Strain repository for the housing of pathogens and other related microorganisms should be developed. It will help in assay development and research work needed to be done for this field.
- Validation of new and existing methods should be done. It should not be limited to the procedures applied in the laboratory; tools used for interpretation of results should also be validated.
- Quality assurance guidelines should be established for microbial forensics laboratories (Budowle 2003).

To develop a proper foundation in the field of microbial forensics, a new group was initiated by Federal Bureau of Investigation known as Scientific Working on Microbial Genetics and Forensics (SWGMGF) on July 29, 2002 (LeBeau 2004). This group provides a common platform for interaction of scientists, academicians from different disciplines and various government agencies for guidelines development in the field of microbial forensics. SWGMGF vision is the development of infrastructure and tools for microbial forensics. The mission of SWGMGF is (Favero et al. 1968):

- defining criteria for development and validation of methods used in forensics for externalisation of biological toxins and microbial agents and,
- define the need for infrastructure development in forensics for active investigation.

As per SWGMGF, following requirements apply for a laboratory routinely involved in microbial forensics work (Budowle 2003):

- Documented procedure for each analytical technique used in the laboratory.
- The procedure should include a proper listing of equipment's, reagents, instructions used at each step, their limitations and literature references.
- The laboratory should have a policy whereby a deviation from an analytical procedure is documented and approved.

4.4.2 Steps in the Investigation of a Suspected Bioterrorism Case

The steps involved in identification of the microorganisms related to a crime scene are explained in Fig. 4.1. The salient aspects that need to be taken into consideration are explained in the following section.

- The suspected case of bio crime with an unusual outbreak of the disease in a place or area.
- Proper collection of samples from the crime scene, which should include every suspected material found at the crime scene. Proper labelling of evidence should be done by mentioning the time and date of collection.
- Samples collected from the crime scene should be considered as potentially biohazardous material, and processing of evidence should be done in a well-equipped laboratory.
- The microbial evidence is collected in the form of live cells, swabs of secretory toxins, DNA swabs, biological samples from the victim (Fluid, skin, hair), clothing and weapons of murder.
- It is imperative that the sample collection is robust, reliable and done sensitively and meticulously. This is because some samples can be very rare and difficult to catch (Bhatia et al. 2015).

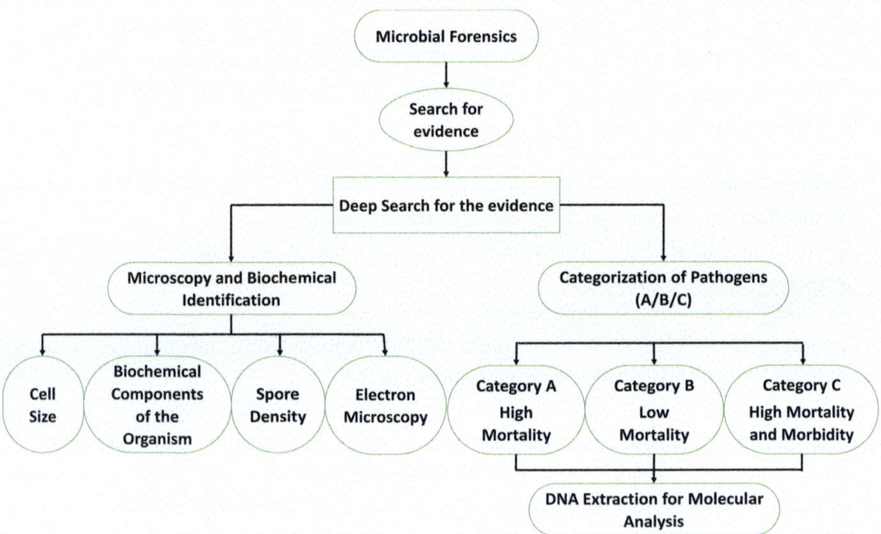

Fig. 4.1 Flow chart showing the steps involved in microbial forensic workflow

4.4.3 Methods Involved in Microbial Forensic Sampling

Correct sampling is quite essential for proper forensic evaluation and successful investigation. Microbial forensics samples can be collected by using three general approaches: bulk collection of an item, collection of a particular portion from an item or collection of liquid and swabbing of the surface (Rose et al. 2004). Bulk collection can be applied to only those items that can be removed easily from the area. Such items are packed appropriately and sent to a laboratory specifically designed to contain hazardous material or toxin. Samples can be collected from different surfaces by swabbing, wiping and vacuuming. It is essential to collect wet swabs, wet wipes and air filter samples throughout the room if bioweapon has been released in a closed confined space. It is also essential to collect the sample from the air vent and air handling units to identify the route of dispersal of the agents (Pattnaik and Sekhar 2008). Usually, sterile rayon swab dipped in phosphate buffer saline solution at pH 7.2 is used to swab a surface horizontally and vertically. From different studies, it has been found that the use of pre-moistened swabs on the porous and non-porous surface is more effective as compared to dry swabs. High-efficiency air particulate vacuum should also be used to collect a sample from the air (Rose et al. 2004). A similar methodology was followed during the investigation of *B. anthracis* contamination and anthrax inhalation case in Washington and New York (Keim et al. 2011). Adhesive tapes have shown better results as compared to pre-moistened swabs on flat porous, non-porous and non-absorbent surfaces (Frawley et al. 2008; Edmonds 2009). Swab collection has been proven more appropriate in case of a small sampling area with a high concentration of the agent, but it has limited value if we compare it with large sampling area. Sample collection form environment is essential for identifying the source of bioweapon agent to any particular geographical location. Apart from this, the sample should be collected from the affected person, his/her workplace and home by a trained medical professional. Use of vacuuming technique along with specialised equipment designed to collect environmental samples and prevent cross-contamination can be considered as an excellent collection method.

4.4.4 Techniques Used in Microbial Forensics

Microbial forensics relies on the DNA-based evidence as well. An important fact to be understood here is that the DNA of convict or victim may not always be available due to sample degradation. In such cases, the DNA of microorganisms associated with biological evidence can be used. The microbial forensics can use the metagenomic DNA from such evidences and analyse the DNA sequences for the purpose of comparing them with the known suspects. A typical workflow of this approach has been given in Fig. 4.2.

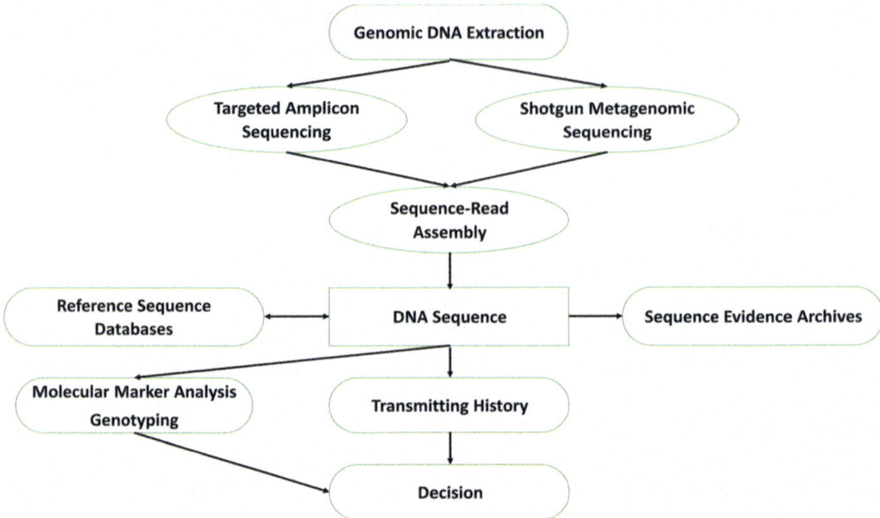

Fig. 4.2 Flow chart showing a typical workflow of molecular marker analysis of DNA and the metagenomic approach

4.4.4.1 MLST

It is a technique that could be used to compare the microorganisms based on the DNA sequence of a group of genes (Maiden et al. 1998). It is a useful method for genetic profiling of bacteria. In this technique, microorganisms are compared based on the "housekeeping" genes that are conserved sequences. Housekeeping genes are those genes that are constitutively expressed at stable levels in the organism as their expression is essential for the functioning of life. Along with the PCR, this technique can be used to amplify specific genetic sequences and can be compared. Reports show the application of MLST in the typing of many bacterial pathogens such as *Neisseria meninigitidis*, *Streptococcus pneumoniae*, among others. MLST data have been employed in epidemiological investigations of various scales and studies of the population biology, pathogenicity, and evolution of bacteria (Maiden 2006). MLST is especially preferred due to good reproducibility of results, the flexibility for analysis of various genes and the ease of analysis of sequencing data. However, the pathogenic strains can be fastidious and can have similar gene sequences. As a result, the sequence variability is less and the significance of results may vary with different sets of genes.

4.4.4.2 PCR-Based Genotyping

Multilocus Variable Number Tandem repeat analysis (MLVA) and Amplified Fragment Length Polymorphism (AFLP) are some of the PCR base typing techniques that are used. AFLP was first described in 1995 (Vos et al. 1995). It is a DNA profiling method combines the PCR and RFLP principles. Bacterial taxonomic research utilizes AFLP because of high-specificity and reproducibility leading to accurate differentiation between bacterial species (Janssen et al. 1996). There are several conserved regions within the closely related pathogens. This is especially significant if the pathogens acquire similar genes through horizontal gene transfer. For such types of pathogen clusters, bi-allelic molecular markers may not be enough for profiling. For closely related groups of pathogens, the VNTR is the marker of choice. VNTRs show greater variability which allows their PCR amplification followed by sequence analysis and semi-quantitative analysis. The size of amplicons may vary based on the number of tandem repeats for the marker sequence (Keim et al. 2000). The VNTR analysis is easily automatable by using fluorescent primers instead of the conventional ones. With the advent of numerous fluorescent tags for DNA, the multiplexing of VNTR analysis is possible. Through multiplexing, a forensic scientist can study multiple loci in the same amplification reaction, hence the name Multilocus VNTR analysis (also known as MLVA) (Klevytska et al. 2001). This approach is successful in differentiating closely related isolates of organisms such as *Bacillus anthracis, Francisella tularensis* and *Yersinia pestis* (Farlow et al. 2001).

4.5 Conclusion

Microbial forensics is a branch of forensic sciences dealing with the microbiological evidence collected from a crime scene. These crime scenes could range from a homicide event up to an act persistent with bioterrorism and biological warfare. The illegitimate use or release of biological agents in the atmosphere or an ecosystem puts the native species at risk, including humans. This is a potential human health hazard and can wipe-out significant proportion of the populations. This makes the establishment of microbial forensics research and diagnostic facilities the need of the hour.

The biological weapons are invisible and are only traceable post-hoc. Microbial forensics allows ad-hoc screening of the various materials for potential microbiota used indiscriminately. Several examples are discussed in this chapter, which only shed light on the potential for microorganisms being used as weapons. The development of bioanalytical and molecular marker-based techniques has come a long way to enable forensic scientists to diagnose or detect the presence of pathogenic microorganisms. For the clinicians, there is need for sensitization and orientation on the management of the infective outbreaks that can potentially lead to pandemics. The need for development of rapid detection mechanisms through molecular-based methods, so that the relevance of microbial forensics is realized and the capacity for containing the outbreaks are built.

References

Aggarwal P, Chopra AK, Gupte S, Sandhu SS (2011) Microbial forensics–an upcoming investigative discipline. J Indian Acad Forensic Med 33:163–165

Baumslag N (2005) Murderous medicine: Nazi doctors, human experimentation, and typhus. Praeger Publishers, Westport

Bhatia M, Mishra B, Thakur A, Dogra V, Loomba PS (2015) Concept of forensic microbiology and its applications

Budowle B (2003) Defining a new forensic discipline: microbial forensics. Profiles DNA 6:7–10

Budowle B, Murch R, Chakraborty R (2005a) Microbial forensics: the next forensic challenge. Int J Legal Med 119:317–330

Budowle B, Schutzer SE, Ascher MS et al (2005b) Toward a system of microbial forensics: from sample collection to interpretation of evidence. Appl Environ Microbiol 71:2209–2213

Christopher LTCGW, Cieslak LTCTJ, Pavlin JA, Eitzen EM (1997) Biological warfare: a historical perspective. Jama 278:412–417

Derbes VJ (1966) De Mussis and the great plague of 1348. JAMA 196:59–62

Edmonds JM (2009) Efficient methods for large-area surface sampling of sites contaminated with pathogenic microorganisms and other hazardous agents: current state, needs, and perspectives. Appl Microbiol Biotechnol 84:811–816

Farlow J, Smith KL, Wong J et al (2001) *Francisella tularensis* strain typing using multiple-locus, variable-number tandem repeat analysis. J Clin Microbiol 39:3186–3192

Favero MS, McDade JJ, Robertsen JA et al (1968) Microbiological sampling of surfaces. J Appl Bacteriol 31:336–343

Franz DR, Jahrling PB, Friedlander AM et al (1997) Clinical recognition and management of patients exposed to biological warfare agents. Jama 278:399–411

Frawley DA, Samaan MN, Bull RL et al (2008) Recovery efficiencies of anthrax spores and ricin from nonporous or nonabsorbent and porous or absorbent surfaces by a variety of sampling methods. J Forensic Sci 53:1102–1107

Greenwood DP (1997) A relative assessment of putative biological-warfare agents. Massachusetts Institute of Technology, Lexington

Jackson PJ, Hugh-Jones ME, Adair DM et al (1998) PCR analysis of tissue samples from the 1979 Sverdlovsk anthrax victims: the presence of multiple *Bacillus anthracis* strains in different victims. Proc Natl Acad Sci 95:1224–1229

Janssen P, Coopman R, Huys G et al (1996) Evaluation of the DNA fingerprinting method AFLP as a new tool in bacterial taxonomy. Microbiology 142:1881–1893

Keim P, Price LB, Klevytska AM et al (2000) Multiple-locus variable-number tandem repeat analysis reveals genetic relationships within *Bacillus anthracis*. J Bacteriol 182:2928–2936

Keim PS, Morse SA, Schutzer SE et al (2011) Microbial forensics, what next? In: Microbial forensics. Elsevier, Amsterdam, pp 693–696

Klevytska AM, Price LB, Schupp JM et al (2001) Identification and characterization of variable-number tandem repeats in the *Yersinia pestis* genome. J Clin Microbiol 39:3179–3185

LeBeau MA (2004) Quality assurance guidelines for laboratories performing forensic analysis of chemical terrorism: scientific working group on forensic analysis of chemical terrorism (SWGFACT). Forensic Sci Commun 6(2):1–14

Leitenberg M (2001) Biological weapons in the twentieth century: a review and analysis. Crit Rev Microbiol 27:267–320

Ligon BL (2006) Plague: a review of its history and potential as a biological weapon. In: Seminars in pediatric infectious diseases. Elsevier, Amsterdam, pp 161–170

Maiden MCJ (2006) Multilocus sequence typing of bacteria. Annu Rev Microbiol 60:561–588

Maiden MCJ, Bygraves JA, Feil E et al (1998) Multilocus sequence typing: a portable approach to the identification of clones within populations of pathogenic microorganisms. Proc Natl Acad Sci 95:3140–3145

Murch RS (2003) Microbial forensics: building a national capacity to investigate bioterrorism. Biosecur Bioterror 1:117–122

National Research Council (US) Panel of Biological Issues (2002) Making the nation safer: The role of science and technology in countering terrorism. National Academies Press, Washington

Paton NI, Leo YS, Zaki SR et al (1999) Outbreak of Nipah-virus infection among abattoir workers in Singapore. Lancet 354:1253–1256

Pattnaik P, Sekhar K (2008) Forensics for tracing microbial signatures: biodefence perspective and preparedness for the unforeseen. https://fas.org/biosecurity/resource/documents/CDC_Bioterrorism_Agents.pdf
Pohanka M, Kuča K (2010) Biological warfare agents. In: Molecular, clinical and environmental toxicology. Springer, Berlin, pp 559–578
Rose L, Jensen B, Peterson A et al (2004) Swab materials and Bacillus anthracis spore recovery from nonporous surfaces. Emerg Infect Dis 10:1023
Stenseth NC, Atshabar BB, Begon M et al (2008) Plague: past, present, and future. PLoS Med 5:e3
Syed A (2018) Nipah virus outbreak in the world. Int J Adv Res Biol Sci 5:131–138
Vos P, Hogers R, Bleeker M et al (1995) A new technique for DNA fingerprinting. Nucleic Acids Res 23:4407–4414
Working R (2001) The trial of Unit 731. In: World commentary. Japan Times. Available via https://www.japantimes.co.jp/opinion/2001/06/05/commentary/world-commentary/the-trial-of-unit-731/. Accessed 26 Apr 2020

Metagenomics Analyses: A Qualitative Assessment Tool for Applications in Forensic Sciences

5

Devika Dileep, Aadya Ramesh, Aarshaa Sojan, Daljeet Singh Dhanjal, Harinder Kaur, and Amandeep Kaur

Abstract

Forensic science deals with the scientific investigation of evidence collected from crime scenes. Since decades, forensic scientists have solved complex and sophisticated crimes using biological evidence. The forensic analyses include the biochemical profiling, physicochemical profiling and molecular analysis of the evidence. The biochemical profiling involves the analysis of the lipid-profile, the presence of toxins in the visceral fluids and enzyme activity analyses. However, it has come to light that detailed molecular evidence, besides the conventional DNA-evidence, is required for the accurate implication of an individual in a crime. Pre-meditated crimes with cadavers found in outbound locations are especially challenging. Many times, the biological samples do not provide enough DNA to carry out the molecular analysis. Metagenomics has paved a way to explore the microbiome of the biological evidence from which the DNA-evidence cannot be extracted. Using high-throughput sequencing methods followed by sequence analysis, the environmental samples of evidence can give valuable insights into the detailed molecular patterns. These patterns can make the identification or comparison of subjects easier. The chapter discusses the genomic analysis, proteome analysis and the metagenomic approaches for investigating a crime scene.

Keywords

DNA profiling · Human microbiota · Metagenomics · Microbial forensics · Molecular markers

D. Dileep · A. Ramesh · A. Sojan · D. S. Dhanjal · H. Kaur · A. Kaur (✉)
School of Bioengineering and Biosciences, Lovely Professional University, Phagwara, Punjab, India

R. S. Chopra et al. (eds.), *Metagenomics: Techniques, Applications, Challenges and Opportunities*, https://doi.org/10.1007/978-981-15-6529-8_5

5.1 Introduction

Forensic science involves the utilization of the laws, principles and tools of science and technology to examine the various physical, chemical and biological evidence around a crime-scene for an amicable solution in the court of law (O'Brien et al. 2015). Nowadays, in cases such as a bioterrorism event, an oil spill or chemical discharge, or a food-borne infectious disease, microbes serve as some of the biological evidence. Analysis of the microbes in such scenarios may be termed as "Microbial Forensics" (Teaf et al. 2018).

Prospecting of the biological evidence at crime scenes should be prioritized, validated and configured using microbial forensics. It plays a significant role in configuring, sample handling and solving the crimes using biological evidence (Budowle et al. 2007). Microbial forensics is a discipline of forensic science that utilizes microbiology to aid in the investigation of biological crimes. It primarily includes the identification and analysis of different microorganisms and their toxins as well as the storage, preparation and mode of delivery of these microbial toxins and pathogens (Pattnaik and Jana 2005).

5.2 Microbes as Weapons

Microbes are potential weapons and not the newly discovered ones, even though less sophisticated. An unfortunate aspect of a biological weapon, when compared to a nuclear weapon, is that it requires much less capital and other investments. Microbes can be cultivated in a small establishment, with meagre investments, and it does not need any elaborate or sophisticated instruments (Thavaselvam and Vijayaraghavan 2010). The need for highly-skilled human resources is also not necessary. Microbes and pathogens can procreate and multiple from a single cell to an uncountable amount if suitable conditions are available. The most common method of using these biological weapons is releasing the virulent pathogens and microbes into the environment that eventually affects the whole community (Riedel 2004).

Any microorganism or microbiological agent can be used as a potential weapon. The factors determining its effectiveness include the ecological stability of the pathogens and the pathogenicity (Cannons et al. 2007). With the advancements in genetic engineering and genetics, common bacteria and organisms can be engineered to produce highly toxic products by inserting ectopically sourced genes. Engineering with transgenes can render these microbes extraordinarily effective and environmentally stable. Biological warfare could be deadly as it uses transferrable and contagious lethal pathogens or toxins to attack the target population. The self-sustainability of these biological attacks can cause severe damage at all levels (van Aken and Hammond 2003).

5.3 Microbes as Environmental Trace Evidence

Different microbes in the environment may serve as trace evidence of a crime committed in that environment. Soil microbiota indicates the evaluation of the degree of decomposition of an excavated/exhumed cadaver and identification of clandestine graves (Hampton-Marcell et al. 2017). Analysis of the microbiome in a water source can assist in determining the extent of pollution of the source or its suitability for human consumption (Poussin et al. 2018). Analysis of the microbes, such as diatoms, is commonly used in cases where a corpse is found in water. This analysis establishes if the drowning was antemortem or postmortem, and if drowning is the cause of death, whether or not the person was killed in the body of water, they were found in (Armstrong and Erskine 2018).

The microbial ecology is dependent on the human microbiome, and the constant interaction between humans and the environment leave specific "microbial signatures" on the human or their accessories (cellphone and shoes). These microbial signatures serve as a means of tracing the person's movements and previous locations (Ursell et al. 2012; Lloyd-Price et al. 2016).

5.3.1 Microbes in Humans

The study of the human microbiome is a dynamic research area. It can yield significant insights into different, forensically relevant aspects such as the identity, geolocation, health and diet of an individual and also to determine the postmortem interval (PMI) and to track the source of infection in case of bio-crimes (Pechal et al. 2018).

Humans harbor microorganisms as part of their physiological development. These microflorae collectively constitute the human microbiome. The normal microbiota can be analyzed to identify a person as the collection of these microflorae vary from person to person and hence serve as a forensic signature (Thursby and Juge 2017). These microflorae are routinely shed, deposited and exchanged continuously, following the principle of exchange given by Locard. Furthermore, since there is a diversity of microbes that are native to different organs of the human body, supporting information like the origin of biological evidence may be determined, that could be used to determine the extent of involvement of humans in crimes (Meadow et al. 2015). An illustration of the expanded horizons of forensic testing using human microbiota is given in Fig. 5.1.

5.3.2 Paleomicrobiology

It refers to the study of ancient microbes and their DNA. Paleo-microbiologists study the epidemiology of ancient infectious diseases and trace the migration patterns of diseases in the past. They use petrified feces (called coprolites) to analyze the gut microbiomes for understanding the dietary aspects of our ancestors (Warinner et al.

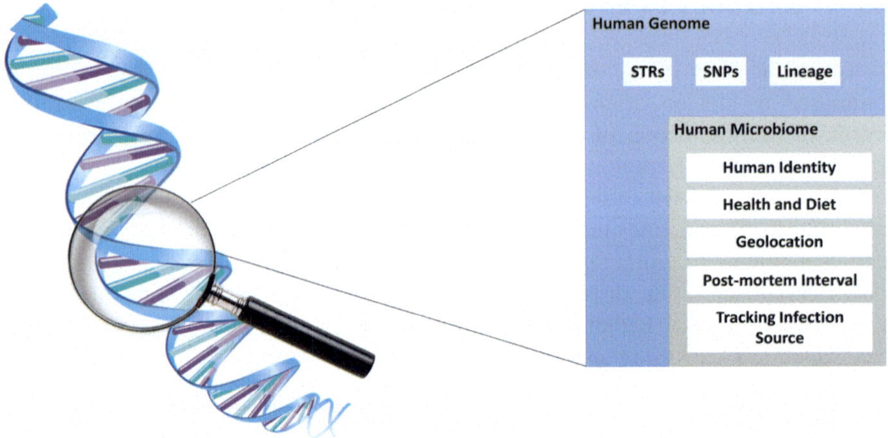

Fig. 5.1 Expanded horizon of forensic testing in investigation

2015). Chas B. Lipman isolated microbes from coal that dated back to millions of years.

Specimens of human origin, have been used by forensic paleo-microbiologists to investigate diseases that have afflicted humankind in the past, including parasitic, viral and bacterial infections (Rivera-Perez et al. 2016).

5.4 Forensic Analysis of Microbes

The Federal Bureau of Investigation (FBI) of the United States of America (USA) has launched the scientific working group on microbial genetics and forensics (SWGMGF), which lays down the infrastructure for microbial forensics (Budowle et al. 2007). The SWGMGF aims to impart knowledge and guidance of the criteria for optimum extraction of information from the available physical evidence. The information can include the identity of the microbial toxins, the convict/victim(s) and the tools used in the execution of a criminal act. The development of a microbial forensics program by SWGMGF enables validation of different methods used in forensic analyses. Apart from the development and validation of methods, the SWGMGF focusses on the quality assurance (QA) as well. QA includes the development of technical proficiencies of the personnel, record-keeping and standard-operating procedures.

DNA profiling is a powerful tool for the identification and elimination of the origin of biological evidence. It may be combined with other effective methods, such as analytical chemistry, pattern matching techniques and microscopy, to give a more accurate characterization of the microbial evidence (Magalhães et al. 2015). A review of the DNA profiling markers has been included later in the chapter.

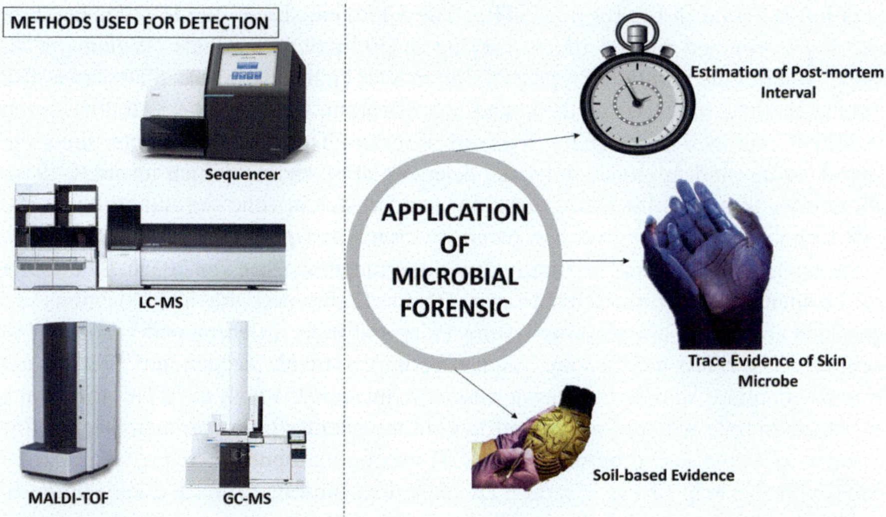

Fig. 5.2 Application of microbial forensics

The isolated microorganisms are sequenced through whole-genome sequencing (WGS). The metagenomes of different niches in the ecosystem are also sequenced and analyzed. The sequencing data then allows the development of the microbial genome databases. These databases enable comparative genomic analysis for forensic analysis of the microbes (Garza and Dutilh 2015). Advanced techniques include the MALDI-TOF-MS method, protein microarrays and DNA, lipid profiling and DNA, expression arrays, SNP method, multiple-locus variable-number tandem repeat analysis (MLVA), confirmative sequencing of PCR product and proteome analyses (analysis of the entire protein profile of a microorganism or cell) (Pattnaik and Jana 2005). These application of these methods in forensic investigation is illustrated in Fig. 5.2. Nucleotide sequencing, along with the comparative evaluation of sequence polymorphism, can determine variations in signature sequences (Pareek et al. 2011).

5.4.1 Human STR

STR stands for Short Tandem Repeats and are also known as microsatellites. These are specific sequences of repetitive nucleotides found in human DNA. The STRs help in the DNA analysis and genetic characterization of individuals based on the differences in the STR patterns (Fan and Chu 2007). DNA analysis has become the go-to identification method for forensic cases where DNA extraction is possible. The human genome has several variations among individuals despite being very similar to each other (Butler 2015). The variations present in the non-coding areas of the human DNA consist of several repeating patterns of nucleotides; one such variation is found in the form of STRs. The STRs consist of one to six nucleotide base pairs repeated several times, varying in different individuals (Willems et al. 2014). The unique number of repeats found in STRs makes an allele. STRs are now a popular method for DNA analysis and identification. FBI uses 13 STR loci in the CODIS (Combined DNA Index System) program. The CODIS program links the central, state and regional forensic science laboratories, which enables these laboratories to search for DNA variations as evidence beyond the state boundaries. STR technology is widely used in paternity tests, other parentage and kinship tests, crime scene investigation and individual identification (Roewer 2013). The STRs can be identified and extracted from a DNA sample after the STR locations are identified the STRs are amplifies using PCR and then differentiated according to their size. The STRs are then analyzed and compared with the standard STRs found from the standard sample in forensic cases (Schneider 2012). If the STRs found in a suspected sample matches with that found in the standard sample, then the identification is positive and otherwise not. STR comparison and matching can also be made with the help of the several STR databases present. STRs are very prone to mutation and approximately make up almost 3% of the human genome (Kayser and De Knijff 2011). The STRs are also prone to polymorphisms. Several developmental and neurological diseases have been associated with STR expansions. STRs also carry out several important functions in humans like the gene expression regulation, DNA repair and duplication, chromatin organization and much more (Press et al. 2014).

5.4.2 DNA and Protein Microarray

Arrays or microarrays are biochips which contain a large number of biological molecules (oligonucleotides for DNA, proteases for proteins, among others) arranged in an array on glass or any other appropriate solid surface. These help to segregate individual molecules from one another and study each molecule independently (Jonczyk et al. 2016).

Microarrays provide information about genome contents with sensitivity, specificity, redundancy, reproducibility and efficiency as required for microbial analysis. Since the 1990s, they have been used for the study of cell biology. It was later used to detect infectious microbial agents (viruses and bacteria) (Miller and Tang, 2009).

A DNA microarray analysis usually follows the steps as under:

- Capture probes are designed from the database of direct sequencing genes of microbes.
- These capture probes are immobilized on the chip (in rRNA analysis, they are usually designed to match the 16S rRNA gene). A single microarray chip can contain up to 400,000 probes.
- Multiplex PCR is carried out of the DNA isolated (labelled with fluorescent dye) from the sample. In some microarrays, PCR is not required. The nucleic acid isolated from the sample is used directly.

- The amplicon (or isolate) is hybridized on to the chip.
- The pattern of hybridization is analyzed using an appropriate imaging mass spectrometry (as per the fluorophore label used for hybridization) and then compared to identify the best match.

Probe lengths may vary from 20–100 nucleotides. Short probes have forensic application and are used in case discrimination of sample using SNP or other short genetic sequences that are unique. Long probes enable detection of related targets (Ziętkiewicz et al. 2012).

Microarrays may be used to determine and measure the host's response to pathogens which further helps to differentiate between strains of a microorganism. The advantage of this technique is that it is even possible to detect species and serotypes that are not specifically represented by the chip; they exhibit unique hybridization signatures (Rasooly and Herold 2008). On the other hand, interference caused by host nucleic acid is a serious setback. While novel sequence may be limitedly detected, its characterization is difficult or impossible, and information on the nature of the variation in the sequence is inadequate (Zhang and Appella 2010).

This technique also lacks provision of the particulars about the genomic context of a feature that has been detected, i.e., whether the gene is encoded in the plasmid or chromosomally or if it originates from horizontal gene transfer and other information about its evolutionary history (Juhas et al. 2009). Pan microbial arrays, like the Greene Chip, were used to detect Plasmodium falciparum, as the cause, during the Marburg virus outbreak. In the SARS outbreak in 2003, VitroChip microarray was helpful for detection (Palacios et al. 2007). Protein arrays work in the similar principle as DNA arrays except that probes are of proteomes of all known strains of a pathogen. The proteomic analysis is especially useful in the detection of engineered strains. The downside is that they are at risk of reduced sensitivity due to changes in the conformation that occur at the time of attachment to the chips (Hall et al. 2007).

5.4.3 DNA and Lipid Profiling

Lipids are majorly present in the membranes of the microorganisms, and some of these are distinctive to a particular microorganism. In environmental microbial studies, the phospholipid ester-linked fatty acids (PLFA) in bacterial membranes are utilized to study bacterial communities (Quideau et al. 2016).

The whole-cell fatty acid analysis is an important tool to characterize a species. It is especially used in the study of the species in genus *Bacillus* due to their extensive use as a bioweapon as a virtue of its pathogenic potential. Fatty acids may be analyzed by FAME (Fatty Acid Methyl Ester) profiling method, which involves the chemical extraction method (Sreenivasulu et al. 2017). In-situ extraction may be done using super-critical fluid for extracting the fatty acids from whole cells or by using pyrolysis in a microreactor, which is then analyzed by FAME. In in-situ extraction, the fatty acids are hydrolyzed and derivatized in a single step, thus saving time. The extracts can be analyzed using different instruments like GC, MS, GC-MS, and HPLC (Gharaibeh and Voorhees 1996).

Lipid biomarkers can be in the form of presence or absence of single or multiple, structurally unique fatty acids. Since they are characteristic to particular genus/species, composite lipid profiles serve as a forensic signature (Abe et al. 2017).

Lipid profiles depend on:

- Type of organism
- Growth conditions
  - Environmental conditions
  - Nutrients medium characteristics

The lipid profile peaks due to growth conditions may be nullified by analyzing the total profile for a lipid specific to an organism or the variation in the relative abundance of the different biomarkers. Hence these methods can also be used for profiling a mixed sample (França et al. 2018).

5.4.3.1 SNP Analysis

The DNA of any organism consists of specific locations which provide the different sequences to various organisms within the same species. These specific locations are known as Single nucleotide polymorphism (SNP). These are the differences in human DNA which are the most common types. There are approximately three million SNPs. Most of the SNPs are considered to be biologically silent as they do not affect the inherited traits or gene function (Vignal et al. 2002). Certain SNPs may affect the gene expression in diseases or might be present in the gene, which affects the function of the protein. SNPs provide a significant ability to understand and provide treatment for human diseases and also provide a genetic marker that helps in identifying and characterizing any species specifically for applications in forensics (Alwi 2005). The stability, frequency and even distribution of the SNPs in the genome provide them with the particular value as genetic markers. An SNP map with high density with the identification of SNP positions on the genome can be used for genetic characterization and speciation of species (Kumar et al. 2012). SNPs serve as genetic markers for genome studies and experiments on fine-scale genetic mapping (Clifford et al. 2004).

5.4.3.2 Multi-Locus VNTR Analysis (MLVA)

In MLVA, multiple target VNTR loci are identified and typed and then compared with a library to identify the species. This analysis, used for microbial species, is comparable to STR typing used for human identification (Pourcel et al. 2011). VNTRs are highly mutable and hence highly discerning, which helps in differentiating between even closely related isolates. Hence when multiple VNTR loci are considered, the discriminating power is greatly intensified (Octavia and Lan 2009).

The process of MLVA may be summarized as follows:

- Isolation of DNA from the target.
- Designing of STR sequence-specific primer sets, as many as required for accurate identification.
- PCR amplification of the target STRs. Size of the PCR products may be determined based on the library of sequenced target organisms.
- Isolation of PCR products by electrophoresis (gel or capillary)
- Comparison of electrogram of PCR products for identification of target species.

The main drawback of MLVA typing for microbial species is the astounding number of species and sub-types which require an equally intensive library of standards for comparison. It would also require a larger number of multi-VNTR, which makes it highly difficult to type all microbial species in this manner (Nadon et al. 2013).

MLVA requires developing specific primers, respective to the identified STR loci and usually requires a large number of primers sets for a single species. (8–25 for *B. anthracis*, 19 for Mycobacterium tuberculosis, 25 for Y. pestis) (Guinard et al. 2017). In the case of an epidemic outbreak, where time is of the essence in identification, the presence of subtypes of species may further complicate the process, making MLVA typing unfeasible (Salaün et al. 2006).

Markers used for MLVA are subject to faster mutation than those used for SNP analysis, making it hard to exclude mismatches completely. This hindrance is further complicated by the fact that the reproducibility of MLVA results between tests done at different labs is a varying measure and each new strain has to be compared with all available results to determine the result (Zaluga et al. 2013).

MLVA can be used in cases of epidemics caused by microbes to determine whether it is naturally caused if it is a case of bioterrorism (anthrax and plague). It was most popularly used in the 2001 anthrax attack in the US. Keim sub typed Bacillus anthracis (used in the 2001 anthrax attack in the US). It linked their presence in clinical samples to samples collected from the patients' food and environment, thereby helping identify the sources of the exposure of the patient to the bacteria (Le Flèche et al. 2001). MLVA used 15 VNTRs to discern 221 genotypes of *B. anthracis* and confirmed the strain that was used as the Ames strain and hence helped dissociate unrelated cases of anthrax. The identification of the strain was also crucial in determining the outbreak as a deliberate attack, as opposed to a natural outbreak, as the Ames strain is a rare strain which is scarcely found in nature (Hoffmaster et al. 2002).

5.4.4 Proteome Analysis

The proteome is the term that usually describes the protein complement to the genome obtained from any organism. Proteomics is defined as the cluster of measurement techniques and methods used for protein analysis (Chandramouli and Qian 2009). Microbial proteomics associates a variety of laboratory methodologies that are based on the following scientific disciplines: biochemistry, analytical chemistry, microbiology and computational sciences. Thus, it is a multidisciplinary scientific field which requires a firm and robust understanding of the limitations, uncertainties and the abilities of the various disciplines (Graves and Haystead 2002). Proteomics identifies the peptides and proteins in a microbial isolate and compares it to other isolates or database of the protein sequence. The aim of the field is sample matching or identification related to organism state or the culture conditions. The various methods utilized for the analysis are gel electrophoresis, mass spectrometry combined with enzymatic digestion and peptide sequencing (Graham et al. 2007).

Proteins provide the expression of genotype, and also the protein complements encoded by any organism provides all the characteristics associated with it. They usually consist of linear chains of 20 different amino acids combined into a particular three-dimensional conformation. Peptides are the smaller chains of amino acids with usually <50 amino acids (Nussinov et al. 2019).

Certain proteins carry out the functions of the bacterial cell, which are division, growth, response to environmental conditions and energy utilization. For specific cellular functions, the proteins involved are produced constitutively. Thus, the identity and sequence of proteins might provide a basis of comparison and organism identification (Merkley et al. 2019).

Mostly microorganisms provide a dynamic response to environmental conditions by changing expressions of various genes. Thus, the proteins expressed by an organism could provide with the details about growth and environment, which provides information for application in forensics (Oonk et al. 2018). However, the challenge is to extract information on what parameters have been used for organism culture based on the given protein profile. A specific protein may have a role in more than one cellular process, and therefore, the expression may be similar to more than one environmental influence (Gräslund et al. 2008). Also, profiles of protein expression can provide a means to differentiate between samples of a single organism cultured under various conditions. The various factors which have been investigated to study their effect on the expression of factors of protein virulence are pH, ionic content and temperature, among others. This information, along with other forensic information, can provide knowledge about the growth history of a collected sample and for comparison purpose (Pérez-Llarena and Bou 2016).

Gel electrophoresis is a separation tool with a low resolution that provides an indication of the size and range of proteins in a collected sample (Zhu et al. 2012). However, it determines only an approximate molecular weight of proteins and is, therefore, challenging to determine the identity of a protein. This problem can be solved by using Mass spectrometry which is a high-resolution analysis tool (Gulcicek et al. 2005).

A mass spectrometer comprises of an inlet system for introducing the sample into the instrument, an ionization source for transferring the analytes as ions into the gas phase and a mass analyzer for detecting the ionized molecules based on the principle of mass-to-charge ratio (m/z) (Rubakhin and Sweedler 2010). In biological mass spectrometry, there are two mechanisms for primary ionization. These are electrospray ionization (ESI), and matrix-assisted laser desorption/ionization (MALDI). Both the techniques are integral and produce the entire molecular ions of biological origin. ESI can provide multiple charges to a molecule and is agreeable to on-line methods of chemical separation. In contrast, the latter provides single charged ions and needs off-line methods of chemical separation (Banerjee and Mazumdar 2012). Also, there are various types of mass analyzers which can be combined with the front-end ionization methods which vary in sensitivity, mass accuracy, resolution and other parameters. In addition, there are mass spectrometers which can carry out tandem experiments for generating more information such as the sequence of the peptides (Han et al. 2008). In these experiments, the initial step is to obtain a spectrum of the entire molecular ions in a given sample. This is followed by various other experiments in which the entire molecular ion is secluded from other components and exposed to molecular fragmentation. This breaks the secluded, entire molecular ion into small pieces which can give a lot of information about the composition of the initial ion present (Liu et al. 2007). A tandem mass spectrometry experiment has the ability to break an entire protein ion into small pieces which indicate the sequence of amino acids that comprise the initial peptide. The masses of proteolytic peptides and related tandem mass spectra can be compared with sequence database already predicted for identification of peptides (DiMaggio and Floudas 2007).

The two approaches for using mass spectrometry for proteomics analysis is usually referred to as top-down and bottom-up proteomics. The goal of both approaches is to identify the proteins in the sample. The latter involves enzymatic digestion of the protein sample such as bacterial isolate to break the proteins into small peptides before mass spectrometric analysis (Zhang et al. 2013). A small portion of the digested sample containing peptide is introduced into a separation apparatus to separate the peptides before mass spectrometric analysis. The individual peptides are identified in the mass spectrometer and then exposed to fragmentation which identifies the sequence of amino acids within each peptide (Gundry et al. 2009). The advantage of the latter is that the small peptide pieces are identified more easily by mass spectrometry with higher specificity and sensitivity. However, complete identification of the digested pieces of the original protein is rare. Thus, important information regarding protein modification (oxidation or phosphorylation) can go undetected and be lost (Angel et al. 2012).

On the other hand, top-down proteomics comprises of performing mass spectrometric analysis on an entire protein followed by tandem mass spectrometric experiments which utilize the process of secondary fragmentation to provide the information related to fragmentation or sequence of the oligopeptide (Lyon et al. 2018). The advantage of this process is the ability to detect proteins of un-sequenced organisms with huge homology to sequenced homology and for determining protein modifications via mass shifts on the intact protein. At the same time, separation of protein can be more challenging than the separation of the peptide. It would be less efficient for large molecules of protein in the gas phase and less sensitive, which would result in less peptide sequence information. Both the approaches have value in applications of bio forensics (Kim et al. 2016).

Hyphenated techniques are available to detect and measure variations in the molecules or proteins of the different strains which aid in pinpointing of its origin and routes of transmission (Patel et al. 2010). These include MALDI-TOF-MS, GC-MS and LC-MS-MS.

5.4.5 MALDI-TOF-MS

It is the abbreviation for matrix-assisted laser disruption ionization-time of flight-mass spectrophotometry. The important uses of this method are rapid microbial identification, direct identification of biomarkers from samples and rapid subtyping. The instrument has a high speed of analysis. The mass spectrophotometer consists of an ionization source, a detector kept under high vacuum and an analyzer. An electrical signal is generated by the ions which are proportional to the number of ions in the detector and a data system documents these signals. It transforms them into a mass spectrum (Clark et al. 2013). The preparation of the sample is comparatively simple. Crude or purified sample is co-crystallized with a high amount of organic matrix onto a target plate which is placed in the mass spectrophotometer. A UV laser is used to produce the ions in most of the microbiological applications (Van Baar 2000). This method could be used directly to cellular suspensions or crude cellular fractions to provide profiles of chemotaxonomic signature. It is also used for the analysis of RNA and DNA, bacteria, bacterial proteomics, rapid characterization of the bacteria at the genus, species and strain level and detection of unknown proteins and their characterization. It can be used for proper identification of light differences between related strains (Singhal et al. 2015).

5.4.6 GC-MS

GC-MS stands for gas chromatography-mass spectrometry. GC-MS is a hyphenated instrumental technique that combines gas chromatography and mass spectrometry. The GC-MS is essentially used to identify different components of a mixture, separate and quantify them. GC-MS is an ideal technique for the identification of elements with low molecular weight. The thermal stability and volatility of a substance are crucial for it to be compatible for analysis by GC-MS. Sample preparation for the technique is unique as materials that can affect the quality of the results may need to be removed (Brattoli et al. 2013). Solvent extraction and different wet chemical methods of sample preparation are also employed. The whole process starts with the gas chromatograph. First, the sample is injected through the inlet, and it gets volatilized. The volatilized sample is effectively separated into its components in the chromatographic column. The carrier gas is the mobile phase and is used to help the volatilized sample to reach the column. The capillary column contains the station phase. The interaction of the sample with the mobile phase and stationary phase results in elution of the sample components at different retention times. Then comes the mass spectrometer, as soon as the sample components leave the column, it gets ionized by the mass spectrometer (Koek et al. 2011). The mass spectrometer uses either a chemical ionization source or an electron ionization source. These different methods have different advantages and help in identifying the molecular weight of components, molecular fingerprints and structural details of the components (Banerjee and Mazumdar 2012). The mass spectrometer has a mass analyzer that separates the molecules that pass through it depending on various mass related properties; the properties analyzed depends on the analyzer used. Common mass analyzers include quadrupole or ion traps. They separate the ions according to the differences in their mass to charge ratios (Clarke 2017). The last step in this technique is the ion detection and analysis, the ions enter a detector, and several peaks are observed. The outcome from the detector is amplified for signal amplification. The signals are stored in the computer and converted to visual displays. The results can be analyzed using various computer libraries and databases and the different components identified (Beale et al. 2018).

5.4.7 LCMS-MS

LC-MS-MS stands for liquid chromatography-tandem mass spectrometry. This powerful analytical tool combines the separating capacity of liquid chromatography with the sensitive analyzing capabilities of tandem mass spectrometry (Pitt 2009). The LC-MS-MS is different from the LC-MS in many significant ways like heightened sensitivity and specificity. Triple quadrupoles are used in the LC-MS-MS system. First, the liquid chromatography is present; separation of components in occurs through the different interaction of the interest samples with the customized mobile phase and stationary phase present (Grebe and Singh 2011). The sample is pumped through the stationary phase using a high pressure liquid mobile phase. After the elution of different components at different retention times due to the different chemical interactions present is directed to the mass spectrometer present. The LC-MS-MS system has a mass spectrometer with an ionization source that nebulizes, de-solvates and ionizes the effluents of the LC (Beccaria and Cabooter 2020). Charged particles are created. Electromagnetic fields help these charged particles to move through a series of mass analyzers under high vacuum. The mass analyzers used are generally quadrupoles. A parent ion or precursor ion with a specific mass to charge ratio is made to pass through the first quadrupole, excluding other particles. The parent ion is converted into daughter ions in the collision cell through fragmentation with an inert gas. Specific daughter ions are targeted using the third quadrupole and those quantified using an electron multiplier. The MS2 present, the fragmentation of parent ions to daughter ions is very specific to the structure of the compound and thus provides heightened selectivity (Girolamo et al. 2013). The MS-MS is 100× times sensitive than MS. The MS-MS targets the interest compound and breaks it down into smaller immediate compounds and filters these products. The filtered products are then identified and detected. This provides for high sensitivity and precision of the LC-MS-MS method. The fragmentation patterns provide more structural information about the component of interest (Sherwood et al. 2009).

The detection protocol should be rapid, effective and specific. High sensitivity in detection methods is required. A molecular diagnosis is the key detection method, but the ambiguity of the microbes used and the partial DNA or unknown DNA of these microbes remains a major setback (Rajapaksha et al. 2019). Prioritizing and identifying biological threats are the key goals of microbial forensics. This can be achieved using several strategies that include creating databases based on identified vulnerable population to provide information, identifying populations that may be vulnerable, developing identification protocols for biological threats using protein signatures, genetic signatures and so on (Casadevall and Relman 2010). The validity of the results and procedures must be constantly updated based present literature. The microbe used as the biological weapon may be unknown; it may not be a human pathogen but maybe a plant pathogen (Pattnaik and Jana 2005). This can cause severe economic damage. Here comes the importance of creating and authenticating a database that could serve as a guide for all the identification procedures and final results. The databases could use inputs from forensics, pure science, genomics and microbiology (Metcalf et al. 2017).

5.5 Role of Metagenomics in Forensic Identification

With the advent of methods in molecular markers as discussed above, genotyping methods and sequencing techniques, the role of metagenomic approaches is receiving much attention. A scene of the crime is full of evidence as there is no perfect crime. The evidence of importance from the metagenomic standpoint includes soil, hair, semen, blood, skin cells and many more. Soil is considered to be ubiquitous due to its presence in forensic evidence. A forensic investigator may find the soil samples in the vehicle tires, shoes, under nails, the skin as well as the discarded murder weapons (Fitzpatrick 2009). Traditionally, the analysis performed on the soil evidence includes the physicochemical analysis, basic microbiological analysis and microscopic analysis. The physicochemical analysis includes analysis of soil texture, color and estimation of the elements present in the soil sample. It is well established that such analysis, although, provides basic analysis, fails to provide an in-depth understanding of the comparison between soils (Moreno et al. 2006). Since the ultimate goal is to match the soil samples, the said matching must be done at the molecular level in detail, so that the evidence is not challenged. On the other hand, the microbiological analysis brings us close to an accurate comparison of the soil samples. However, it is limited by the fact that the number of culturable microbial species in the soil sample is only around 10%. Therefore, there is a need for bringing metagenomic analysis to cater to the need for complete microbial diversity analysis. Metagenomics is the study of the genetic composition of all microbial species in an environmental sample. These samples may include the soil, water, sediments, among others. Since soil contains a rich population of the microorganisms, the sheer diversity of the microbes makes the comparison of soil samples difficult. The method of choice used for elucidation of the microbial diversity of a soil sample is highlighted in Fig. 5.3. The basic idea is to use the metagenomic sequencing

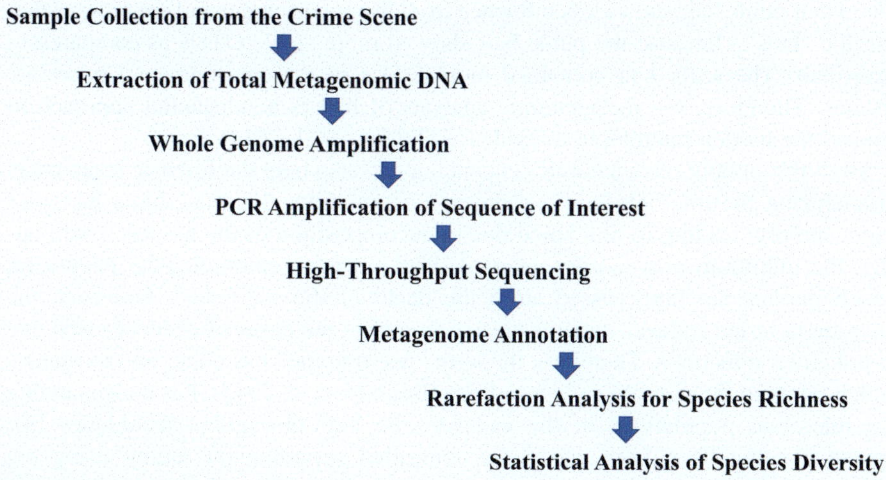

Fig. 5.3 Flow chart showing the workflow of forensic metagenomic analysis of soil samples

approach for obtaining the reads from the environmental samples. The reads are then annotated using online tools such as MG-RAST (Metagenomic Rapid Annotations using Subsystems Technology) (Khodakova et al. 2014).

Similar to the soil, another potential evidence at the crime scene is the hair samples. Both scalp hair and pubic hair are acceptable evidence in the court of law. However, the comparison is usually made using the microscopic examination and hair-pigmentation (Houck 2005). Some success in fruitful utilization of hair as biological evidence is due to the molecular analysis. Techniques such as STR profiling can prospect the genome for unique gene signatures for identifying a person of interest (Linch et al. 1998). For the hair sample, the molecular marker analysis is possible only when the hair follicle is intact for extracting DNA. Lack of sufficient intact chromosomal DNA is a persistent challenge for most hair-based evidence. The lacuna is overcome using the metagenomic approach, in which culture-independent analysis of the hair-microbiota is performed. In one such study, the subjects collected the scalp and pubic hair samples using the prescribed collection kits and tools. The forensic team cut the hair samples into short fragments and extracted the total metagenomic DNA from the hair sample. The method followed in here was a direct DNA extraction method. The team then amplified the 16S rRNA gene using specific primers containing a sequencing adapter and multiplex identifier (MID) tag sequence. Unique MID tag sequences were used for the scalp and pubic hair from males and females. The amplicons were ~350 base pairs long and sequenced. The sequence reads that were obtained in this study were pre-processed using the QIIME (Quantitative Insights into Microbial Ecology). The QIIME output was loaded on the USEARCH61 for eliminating the overlapping sequences and assembling the operational taxonomic units (OTUs). The clustering analysis of the OTUs was done using the principal coordinate analysis plot (PCoA) method. Ultimately, the OTUs of the cohabiting subjects were compared with the non-cohabiting subjects. The research found that pubic hair shows less influence from the environment and are more niche-specific. This is because the pubic hair showed more stable OTUs as compared to scalp hair. The scalp hair contained more OTUs of the environmental microbial species. Therefore, the metagenomic analysis of hair is a substantial approach to unravel the hidden microbiota as evidence (Tridico et al. 2014).

High-throughput metagenomic sequencing can also help the forensic researchers in analyzing the microbiota of a cadaver. Post-mortem, this microbiota increases exponentially, leading to foul smell due to decomposition of the tissues. Traditionally, the minimum post-mortem interval (PMI_{min}) is an entomological parameter, which implies the time passed since the death of the individual. However, the microbiota in the cadaver remains elusive except in the cases of deaths caused due to unknown infections. Therefore, the term "necrobiome" has come into existence, which refers to the microbiota in a cadaver (Benbow et al. 2013). For understanding the microbial populations in the cadaver, the high-throughput techniques like pyrosequencing have been used in a controlled experimental setting using pig cadavers. In the research, the pig carcass swabs were collected from various sites including buccal cavity, tongue and skin. The metagenomic DNA was extracted from the swabs and amplified using specific tagged 16S rRNA primers. The amplicons were cloned to prepare a metagenomic library. The pooled clones from the library were sequenced, and the sequencing reads were processed. In the processing, the short reads, non-ribosomal DNA sequences and the primer sequences were eliminated using the black box chimera check (B2C2) program. The sequences after processing were aligned using the infernal aligner tool of the Ribosomal Database Project (RDP) to generate the final sequences. From the complete sequence data, the diversity was correlated with the time of decomposition after death. Using the correlation data, the relationship between the species-richness and the time showed the progression of decomposition (Pechal et al. 2014).

5.6 Conclusion

Microorganisms are at the forefront of forensic investigation as biological evidence. Especially the human microbiota provides a reliable signature when analyzed from exhumed cadavers or a water-borne corpse. The analysis of the latter is especially useful in drowning cases. The microorganisms in the fossilized remains provide vital information during characterization of such fossils. The forensic analysis of microorganisms is carried out using molecular markers through DNA profiling. High throughput DNA sequencing can help in the identification of microbial diversity at the level of single nucleotides. Various genotyping markers such as SNPs, MLVs and STRs (minisatellites) provide detailed molecular analysis of genetic variations among different forensic samples. Apart from the DNA based markers, the biochemical profiling for lipids and the proteome profile also enable a targeted analysis of the evidence. However, for DNA analysis to be possible, the biological evidence should contain sufficient intact nuclear material. For trace evidence such as hair and soil, DNA extraction is a challenge. Therefore, the metagenomic analysis of the microbial species associated with such samples provides the molecular signature the forensic experts look for. Metagenomic analysis of the soil and hair samples follows many similar strategies. However, the quality and quantity of DNA analysis involved in these samples are overwhelming. Metagenomic analysis of the soil and hair samples require correlation with the already existing microscopic, physico-chemical and biochemical analysis. There is also scope for standardization of forensic procedures for developing standalone methods of forensic analysis. Overall, metagenomic analysis is highly recommendable for inclusion into mainstream forensic analyses.

References

Abe H, Yajima D, Hoshioka Y et al (2017) Myoglobinemia markers with potential applications in forensic sample analysis: lipid markers in myoglobinemia for postmortem blood. Int J Legal Med 131:1739–1746. https://doi.org/10.1007/s00414-017-1657-8

Alwi ZB (2005) The use of SNPs in pharmacogenomics studies. Malays J Med Sci 12:4

Angel TE, Aryal UK, Hengel SM et al (2012) Mass spectrometry-based proteomics: existing capabilities and future directions. Chem Soc Rev 41:3912–3928

Armstrong EJ, Erskine KL (2018) Investigation of drowning deaths: a practical review. Acad Forensic Pathol 8:8–43

Banerjee S, Mazumdar S (2012) Electrospray ionization mass spectrometry: a technique to access the information beyond the molecular weight of the analyte. Int J Anal Chem 2012:1–40. https://doi.org/10.1155/2012/282574

Beale DJ, Pinu FR, Kouremenos KA et al (2018) Review of recent developments in GC–MS approaches to metabolomics-based research. Metabolomics 14:1–31

Beccaria M, Cabooter D (2020) Current developments in LC-MS for pharmaceutical analysis. Analyst 145:1129–1157

Benbow ME, Lewis AJ, Tomberlin JK, Pechal JL (2013) Seasonal necrophagous insect community assembly during vertebrate carrion decomposition. J Med Entomol 50:440–450

Brattoli M, Cisternino E, Rosario Dambruoso P et al (2013) Gas chromatography analysis with olfactometric detection (GC-O) as a useful methodology for chemical characterization of odorous compounds. Sensors (Switzerland) 13:16759–16800

Budowle B, Beaudry JA, Barnaby NG et al (2007) Role of law enforcement response and microbial forensics in investigation of bioterrorism. Croat Med J 48:437–449

Butler JM (2015) The future of forensic DNA analysis. Philos Trans R Soc B Biol Sci 370:20140252. https://doi.org/10.1098/rstb.2014.0252

Cannons A, Amuso P, Anderson B (2007) Biotechnology and the public health response to bioterrorism. In: Microorganisms and bioterrorism. Springer, Boston, pp 1–13

Casadevall A, Relman DA (2010) Microbial threat lists: obstacles in the quest for biosecurity? Nat Rev Microbiol 8:149–154

Chandramouli K, Qian P-Y (2009) Proteomics: challenges, techniques and possibilities to overcome biological sample complexity. Hum Genomics Proteomics 2009:239204. https://doi.org/10.4061/2009/239204

Clark AE, Kaleta EJ, Arora A, Wolk DM (2013) Matrix-assisted laser desorption ionization-time of flight mass spectrometry: a fundamental shift in the routine practice of clinical microbiology. Clin Microbiol Rev 26:547–603. https://doi.org/10.1128/CMR.00072-12

Clarke W (2017) Mass spectrometry in the clinical laboratory: determining the need and avoiding pitfalls. In: Mass spectrometry for the clinical laboratory. Elsevier, Amsterdam, pp 1–15

Clifford RJ, Edmonson MN, Nguyen C et al (2004) Bioinformatics tools for single nucleotide polymorphism discovery and analysis. Ann N Y Acad Sci 1020:101–109

DiMaggio PA, Floudas CA (2007) De novo peptide identification via tandem mass spectrometry and integer linear optimization. Anal Chem 79:1433–1446. https://doi.org/10.1021/ac0618425

Fan H, Chu JY (2007) A brief review of short tandem repeat mutation. Genomics Proteomics Bioinformatics 5:7–14

Fitzpatrick RW (2009) Soil: forensic analysis. Wiley Encycl Forensic Sci:1–14. https://doi.org/10.1002/9780470061589.fsa096

França CN, Mendes CC, Ferreira CES (2018) Time collection and storage conditions of lipid profile. Braz J Med Biol Res 51:e6955. https://doi.org/10.1590/1414-431X20176955

Garza DR, Dutilh BE (2015) From cultured to uncultured genome sequences: Metagenomics and modeling microbial ecosystems. Cell Mol Life Sci 72:4287–4308

Gharaibeh AA, Voorhees KJ (1996) Characterization of lipid fatty in whole-cell microorganisms using in situ supercritical fluid derivatization/extraction and chromatography/mass spectrometry. Anal Chem 68:2805–2810. https://doi.org/10.1021/ac9600767

Girolamo F, Lante I, Muraca M, Putignani L (2013) The role of mass spectrometry in the "omics" era. Curr Org Chem 17:2891–2905. https://doi.org/10.2174/13852728178881311l8162725

Graham RLJ, Graham C, McMullan G (2007) Microbial proteomics: a mass spectrometry primer for biologists. Microb Cell Factories 6:26

Gräslund S, Nordlund P, Weigelt J et al (2008) Protein production and purification. Nat Methods 5:135–146

Graves PR, Haystead TAJ (2002) Molecular biologist's guide to proteomics. Microbiol Mol Biol Rev 66:39–63. https://doi.org/10.1128/mmbr.66.1.39-63.2002

Grebe SKG, Singh RJ (2011) LC-MS/MS in the clinical laboratory - where to from here? Clin Biochem Rev 32:5–31

Guinard J, Latreille A, Guérin F et al (2017) New multilocus variable-number tandem-repeat analysis (MLVA) scheme for fine-scale monitoring and microevolution-related study of *Ralstonia pseudosolanacearum* phylotype I populations. Appl Environ Microbiol 83:e03095-16. https://doi.org/10.1128/AEM.03095-16

Gulcicek EE, Colangelo CM, McMurray W et al (2005) Proteomics and the analysis of proteomic data: an overview of current protein-profiling technologies. Curr Protoc Bioinformatics 10:13.1.1–13.1.31. https://doi.org/10.1002/0471250953.bi1301s10

Gundry RL, White MY, Murray CI et al (2009) Preparation of proteins and peptides for mass spectrometry analysis in a bottom-up proteomics workflow. Curr Protoc Mol Biol Chapter 10: Unit10.25. https://doi.org/10.1002/0471142727.mb1025s88

Hall DA, Ptacek J, Snyder M (2007) Protein microarray technology. Mech Ageing Dev 128:161–167. https://doi.org/10.1016/j.mad.2006.11.021

Hampton-Marcell JT, Lopez JV, Gilbert JA (2017) The human microbiome: an emerging tool in forensics. Microb Biotechnol 10:228–230

Han X, Aslanian A, Yates JR (2008) Mass spectrometry for proteomics. Curr Opin Chem Biol 12:483–490

Hoffmaster AR, Fitzgerald CC, Ribot E et al (2002) Molecular subtyping of Bacillus anthracis and the 2001 bioterrorism-associated anthrax outbreak, United States. Emerg Infect Dis 8:1111–1116. https://doi.org/10.3201/eid0810.020394

Houck MM (2005) Forensic human hair examination and comparison in the 21st century. Forensic Sci Rev 17:51–66

Jonczyk R, Kurth T, Lavrentieva A et al (2016) Living cell microarrays: an overview of concepts. Microarrays 5:11. https://doi.org/10.3390/microarrays5020011

Juhas M, Van Der Meer JR, Gaillard M et al (2009) Genomic islands: tools of bacterial horizontal gene transfer and evolution. FEMS Microbiol Rev 33:376–393

Kayser M, De Knijff P (2011) Improving human forensics through advances in genetics, genomics and molecular biology. Nat Rev Genet 12:179–192

Khodakova AS, Smith RJ, Burgoyne L et al (2014) Random whole metagenomic sequencing for forensic discrimination of soils. PLoS One 9:e104996

Kim MS, Zhong J, Pandey A (2016) Common errors in mass spectrometry-based analysis of post-translational modifications. Proteomics 16:700–714

Koek MM, Jellema RH, van der Greef J et al (2011) Quantitative metabolomics based on gas chromatography mass spectrometry: status and perspectives. Metabolomics 7:307–328

Kumar S, Banks TW, Cloutier S (2012) SNP discovery through next-generation sequencing and its applications. Int J Plant Genomics 2012:831460. https://doi.org/10.1155/2012/831460

Le Flèche P, Hauck Y, Onteniente L et al (2001) A tandem repeats database for bacterial genomes: application to the genotyping of *Yersinia pestis* and *Bacillus anthracis*. BMC Microbiol 1:1–14. https://doi.org/10.1186/1471-2180-1-2

Linch CA, Smith SL, Prahlow JA (1998) Evaluation of the human hair root for DNA typing subsequent to microscopic comparison. J Forensic Sci 43:305–314

Liu T, Belov ME, Jaitly N et al (2007) Accurate mass measurements in proteomics. Chem Rev 107:3621–3653

Lloyd-Price J, Abu-Ali G, Huttenhower C (2016) The healthy human microbiome. Genome Med 8:51

Lyon YA, Riggs D, Fornelli L et al (2018) The ups and downs of repeated cleavage and internal fragment production in top-down proteomics. J Am Soc Mass Spectrom 29:150–157. https://doi.org/10.1007/s13361-017-1823-8

Magalhães T, Dinis-Oliveira RJ, Silva B et al (2015) Biological evidence management for DNA analysis in cases of sexual assault. Sci World J 2015:365674. https://doi.org/10.1155/2015/365674

Meadow JF, Altrichter AE, Bateman AC et al (2015) Humans differ in their personal microbial cloud. PeerJ 3:e1258. https://doi.org/10.7717/peerj.1258

Merkley ED, Wunschel DS, Wahl KL, Jarman KH (2019) Applications and challenges of forensic proteomics. Forensic Sci Int 297:350–363

Metcalf JL, Xu ZZ, Bouslimani A et al (2017) Microbiome tools for forensic science. Trends Biotechnol 35:814–823

Miller MB, Tang YW (2009) Basic concepts of microarrays and potential applications in clinical microbiology. Clin Microbiol Rev 22:611–633

Moreno LI, Mills DK, Entry J et al (2006) Microbial metagenome profiling using amplicon length heterogeneity-polymerase chain reaction proves more effective than elemental analysis in discriminating soil specimens. J Forensic Sci 51:1315–1322

Nadon CA, Trees E, Ng LK et al (2013) Development and application of MLVA methods as a tool for inter-laboratory surveillance. Eur Secur 18:20565. https://doi.org/10.2807/1560-7917.ES2013.18.35.20565

Nussinov R, Tsai CJ, Jang H (2019) Protein ensembles link genotype to phenotype. PLoS Comput Biol 15:e1006648

O'Brien É, Daeid NN, Black S (2015) Science in the court: pitfalls, challenges and solutions. Philos Trans R Soc B Biol Sci 370:20150062. https://doi.org/10.1098/rstb.2015.0062

Octavia S, Lan R (2009) Multiple-locus variable-number tandem-repeat analysis of *Salmonella enterica serovar Typhi*. J Clin Microbiol 47:2369–2376. https://doi.org/10.1128/JCM.00223-09

Oonk S, Schuurmans T, Pabst M et al (2018) Proteomics as a new tool to study fingermark ageing in forensics. Sci Rep 8:1–11. https://doi.org/10.1038/s41598-018-34791-z

Palacios G, Quan PL, Jabado OJ et al (2007) Panmicrobial oligonucleotide array for diagnosis of infectious diseases. Emerg Infect Dis 13:73–81. https://doi.org/10.3201/eid1301.060837

Pareek CS, Smoczynski R, Tretyn A (2011) Sequencing technologies and genome sequencing. J Appl Genet 52:413–435

Patel K, Patel J, Patel M et al (2010) Introduction to hyphenated techniques and their applications in pharmacy. Pharm Methods 1:2. https://doi.org/10.4103/2229-4708.72222

Pattnaik P, Jana AM (2005) Microbial forensics: applications in bioterrorism. Environ Forensic 6:197–204

Pechal JL, Crippen TL, Benbow ME et al (2014) The potential use of bacterial community succession in forensics as described by high throughput metagenomic sequencing. Int J Legal Med 128:193–205

Pechal JL, Schmidt CJ, Jordan HR, Benbow ME (2018) A large-scale survey of the postmortem human microbiome, and its potential to provide insight into the living health condition. Sci Rep 8:5724. https://doi.org/10.1038/s41598-018-23989-w

Pérez-Llarena FJ, Bou G (2016) Proteomics as a tool for studying bacterial virulence and antimicrobial resistance. Front Microbiol 7:410

Pitt JJ (2009) Principles and applications of liquid chromatography-mass spectrometry in clinical biochemistry. Clin Biochem Rev 30:19–34

Pourcel C, Minandri F, Hauck Y et al (2011) Identification of variable-number tandem-repeat (VNTR) sequences in *Acinetobacter baumannii* and interlaboratory validation of an optimized multiple-locus VNTR analysis typing scheme. J Clin Microbiol 49:539–548. https://doi.org/10.1128/JCM.02003-10

Poussin C, Sierro N, Boué S et al (2018) Interrogating the microbiome: experimental and computational considerations in support of study reproducibility. Drug Discov Today 23:1644–1657

Press MO, Carlson KD, Queitsch C (2014) The overdue promise of short tandem repeat variation for heritability. Trends Genet 30:504–512

Quideau SA, McIntosh ACS, Norris CE et al (2016) Extraction and analysis of microbial phospholipid fatty acids in soils. J Vis Exp 2016(114):e54360. https://doi.org/10.3791/54360

Rajapaksha P, Elbourne A, Gangadoo S et al (2019) A review of methods for the detection of pathogenic microorganisms. Analyst 144:396–411

Rasooly A, Herold KE (2008) Food microbial pathogen detection and analysis using DNA microarray technologies. Foodborne Pathog Dis 5:531–550

Riedel S (2004) Biological warfare and bioterrorism: a historical review. Bayl Univ Med Cent Proc 17:400–406. https://doi.org/10.1080/08998280.2004.11928002

Rivera-Perez JI, Santiago-Rodriguez TM, Toranzos GA (2016) Paleomicrobiology: a snapshot of ancient microbes and approaches to forensic microbiology. Microbiol Spectr 4:EMF-0006-2015. https://doi.org/10.1128/microbiolspec.emf-0006-2015

Roewer L (2013) DNA fingerprinting in forensics: past, present, future. Investig Genet 4:22

Rubakhin SS, Sweedler JV (2010) A mass spectrometry primer for mass spectrometry imaging. Methods Mol Biol 656:21–49. https://doi.org/10.1007/978-1-60761-746-4_2

Salaün L, Mérien F, Gurianova S et al (2006) Application of multilocus variable-number tandem-repeat analysis for molecular typing of the agent of leptospirosis. J Clin Microbiol 44:3954–3962. https://doi.org/10.1128/JCM.00336-06

Schneider PM (2012) Beyond STRs: the role of diallelic markers in forensic genetics. Transfus Med Hemother 39:176–180

Sherwood CA, Eastham A, Lee LW et al (2009) Rapid optimization of MRM-MS instrument parameters by subtle alteration of precursor and product m/z targets. J Proteome Res 8:3746–3751. https://doi.org/10.1021/pr801122b

Singhal N, Kumar M, Kanaujia PK, Virdi JS (2015) MALDI-TOF mass spectrometry: an emerging technology for microbial identification and diagnosis. Front Microbiol 6:791

Sreenivasulu B, Paramageetham C, Sreenivasulu D et al (2017) Analysis of chemical signatures of alkaliphiles using fatty acid methyl ester analysis. J Pharm Bioallied Sci 9:106–114. https://doi.org/10.4103/jpbs.JPBS_286_16

Teaf CM, Flores D, Garber M, Harwood VJ (2018) Toward forensic uses of microbial source tracking. Microbiol Spectr 6:EMF-0014-2017. https://doi.org/10.1128/microbiolspec.emf-0014-2017

Thavaselvam D, Vijayaraghavan R (2010) Biological warfare agents. J Pharm Bioallied Sci 2:179. https://doi.org/10.4103/0975-7406.68499

Thursby E, Juge N (2017) Introduction to the human gut microbiota. Biochem J 474:1823–1836

Tridico SR, Murray DC, Addison J et al (2014) Metagenomic analyses of bacteria on human hairs: a qualitative assessment for applications in forensic science. Investig Genet 5:16

Ursell LK, Metcalf JL, Parfrey LW, Knight R (2012) Defining the human microbiome. Nutr Rev 70: S38. https://doi.org/10.1111/j.1753-4887.2012.00493.x

van Aken J, Hammond E (2003) Genetic engineering and biological weapons. New technologies, desires and threats from biological research. EMBO Rep 4:S57. https://doi.org/10.1038/sj.embor.embor860

Van Baar BLM (2000) Characterisation of bacteria by matrix-assisted laser desorption/ionisation and electrospray mass spectrometry. FEMS Microbiol Rev 24:193–219

Vignal A, Milan D, SanCristobal M, Eggen A (2002) A review on SNP and other types of molecular markers and their use in animal genetics. Genet Sel Evol 34:275. https://doi.org/10.1186/1297-9686-34-3-275

Warinner C, Speller C, Collins MJ, Lewis CM (2015) Ancient human microbiomes. J Hum Evol 79:125–136. https://doi.org/10.1016/j.jhevol.2014.10.016

Willems T, Gymrek M, Highnam G et al (2014) The landscape of human STR variation. Genome Res 24:1894–1904. https://doi.org/10.1101/gr.177774.114

Zaluga J, Stragier P, Van Vaerenbergh J et al (2013) Multilocus variable-number-tandem-repeats analysis (MLVA) distinguishes a clonal complex of *Clavibacter michiganensis subsp. michiganensis* strains isolated from recent outbreaks of bacterial wilt and canker in Belgium. BMC Microbiol 13:126. https://doi.org/10.1186/1471-2180-13-126

Zhang N, Appella DH (2010) Advantages of peptide nucleic acids as diagnostic platforms for detection of nucleic acids in resource-limited settings. J Infect Dis 201:S42–S45. https://doi.org/10.1086/650389

Zhang Y, Fonslow BR, Shan B et al (2013) Protein analysis by shotgun/bottom-up proteomics. Chem Rev 113:2343–2394

Zhu Z, Lu JJ, Liu S (2012) Protein separation by capillary gel electrophoresis: a review. Anal Chim Acta 709:21–31

Ziętkiewicz E, Witt M, Daca P et al (2012) Current genetic methodologies in the identification of disaster victims and in forensic analysis. J Appl Genet 53:41–60

Realizing Bioremediation Through Metagenomics: A Technical Review

6

Deepansh Sharma, Deepti Singh, Mehak Manzoor, Kunal Meena, Vikrant Sharma, Kajal Butaney, and Reshan Gale Marbaniang

Abstract

A variety of toxic pollutants have been reported in industrial effluents, which roots grave environmental pollution. Conventional culture-based strategies have led to isolation, screening and identification of potential microorganisms for bioremediation. Microscopic analysis and molecular marker studies establish that less than 1% of the microorganisms can be cultured using routine microbiological techniques. The recent progress in culture-independent approach has laid down into the ecology of environmental sites. Because of the various constraints related to the culture-methods, a strategy identified as "metagenomics" is now preferred to discover the genetic makeup of both culturable and non-culturable microbes from any sample. In the present book chapter, we have discussed scope of the metagenomic approaches for utilisation in environmental clean-up of affected sites. Metagenomic tools, composed with developing high-throughput sequencing knowledges have elaborated understanding into the complex exchanges among diverse microbial populations and their metabolic ability, which could be utilized in different bioremediation approaches for effective exclusion contaminants from the environment.

Keywords

Bioremediation · Culture-independent methods · Degradative genes · Functional metagenomics

D. Sharma · D. Singh (✉) · M. Manzoor · K. Meena · V. Sharma · K. Butaney · R. G. Marbaniang
Amity Institute of Microbial Technology, Amity University Rajasthan, Jaipur, India

R. S. Chopra et al. (eds.), *Metagenomics: Techniques, Applications, Challenges and Opportunities*, https://doi.org/10.1007/978-981-15-6529-8_6

6.1 Introduction

Industrialization and urbanisation are fuelling the need for a proper understanding of the association and impact of microbial contaminants and human health. In many developing countries, the rising pollution and accumulation of pollutants are the result of the fast industrialisation in the last few decades. A variety of toxic pollutants have been reported in industrial effluents, which roots grave environmental pollution (Saharan et al. 2011). The recalcitrant nature and poor biodegradability of organic and inorganic contaminants in industrial effluent have a crucial impact for environmental well-being protection and, thus, there is an immediate need to tackle such kind of industrial effluent sufficiently before the disposal.

The traditional approaches for waste biodegradation are composting, landfilling, dumping sites, incineration of waste by heat and chemical neutralization of the various contaminants. These approaches are not accepted worldwide due to their intricate design and lack of public acceptance due to non-ethical presentation (Karigar and Rao 2011; Kumar et al. 2015). Microbial-mediated degradation of the various pollutants is a practical approach to bioremediate the pollutants. Microbial bioremediation is known as an amicable, economically feasible and eco-friendly alternative that delivers bearable ways to clean up of the pollutants (Sharma and Dhanjal 2016; Sharma et al. 2017; Vidali 2001; Leung 2004; Sharma 2016).

Microbial degradation of the pollutants involves the process of detoxification and mineralisation. These processes ultimately convert the pollutants into simpler molecules like carbon dioxide, water, and methane. Though, if the pollutants are obstinate in an environment, their biodegradation is often regulated at multiple levels by exploiting various enzymes or microbial communities. The action of microbial degradation primarily is contingent upon the metabolic profile of microbial cells to detoxify the accumulating pollutants, which mainly rely on the type of pollutants, availability and ability of the microbe to depolymerise it (Antizar-Ladislao 2010).

However, every so often, the microbial bioremediation of industrial pollutants occurs at a slow pace because of the inability of the indigenous microbiome to detoxify. Sometimes to tackle such problems, various potential microbes and their efficient enzymes are augmented to the system for improved degradability. Furthermore, this strategy may be invasive or foreign to the native microbiome, which can continue long after the pollutant has been degraded (Techtmann and Hazen 2016). Conventional culture-based strategies have led to isolation, screening and identification of potential microorganisms for bioremediation. Microscopic analysis and molecular marker studies establish that less than 1% of the microorganisms can be cultured using routine microbiological techniques. So, over 99% microbes in the samples collected from the environment are not readily available for conventional research (Handelsman 2004). The recent progress in culture-independent approach has laid down into the ecology of environmental sites (Gadd 2010). The conventional culture-based approach has provided only inadequate evidence about the native microbial communities of contaminated sites.

The recognition and listing of microbial communities by using traditional 16S rDNA gene manipulation and sequencing represent only the major microbial phyla dominant in the polluted sites. The triumph of microbial bioremediation downplays for reasons like absence of information on the microbial population dominant in the polluted sites; inadequate information on the metabolic pathways, the aspects of monitoring the survival and metabolic activity of microbial communities; and the changes in environmental parameters (Lovley 2003). Because of the various constraints related to the culture-methods, a strategy identified as "metagenomics" is now preferred to discover the genetic makeup of both culturable and non-culturable microbes from any sample (Handelsman et al. 1998).

In the metagenomic approach, a pool of genomes extracted from a contaminated sample is utilized to recognize the genes intricated in bioremediation. Whole genomes are extracted directly from the environmental samples and the desired genes are isolated, which increases the efficiency of the approach (Yergeau et al. 2012). With the progress in the research related to the vector selection and construction, now it is possible to work efficiently on large genomic fragments and later screen large clone libraries with functional metagenome (Lorenz and Eck 2005). In the present book chapter, the scope of the metagenomic approaches for utilisation in environmental clean-up of affected sites.

6.2 Metagenome/Culture-Independent Approach

The science of metagenomics is still new, and the microbiomes of different ecological niches are mostly unexplored because of the large microbial communities and their interactions, and the non-availability of their nutritional requirements. As a result, we cannot detect most microbial species using conventional culture-based methods (Riesenfeld et al. 2004). The idea behind metagenomics presented by Handelsman et al. (1998) includes the mining of the metagenomic DNA of all the microorganisms of an environmental sample for constructing a metagenomic library and subsequent screening of the created metagenomic library. Initially, the main objective of the approach was to identify the unculturable populations that explicitly catalyse the biological activities and recognise the genes related to them. This gave rise to the term functional metagenome (Fig. 6.1) (Uchiyama and Miyazaki 2009; Venter et al. 2004). Apart from the functional metagenomics, various other approaches such as the shotgun approach, meta-proteomics, DNA microarray and phylogenetic are commonly practised to study microbial communities of environmental sites.

6.3 The Metagenomic Approach: Retrieving Biodegradation Genes

As an example, here we discuss the metagenomic approach for discovering the novel enzyme-coding genes for applications in biodegradation. Microbial degradation of any polymeric substrate requires the knowledge of microbial physiology. These polymers can be polysaccharides for the production of bioethanol or the plastics in

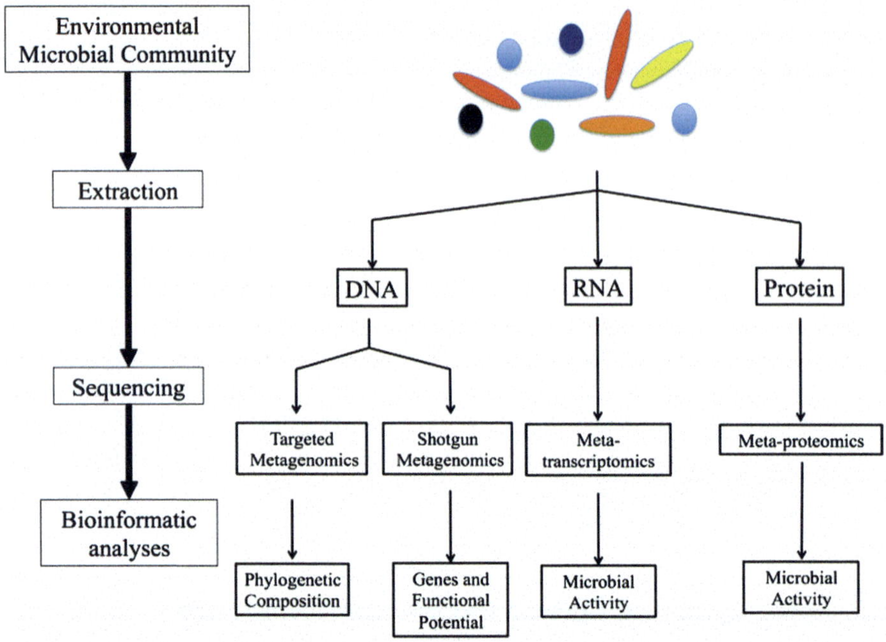

Fig. 6.1 Metagenomic approaches to understanding microbial communities (Courtesy: Techtmann and Hazen 2016)

case of environmental remediation. In any case, the limiting-step is the screening of the metagenomic library for the presence of the genes with the potential to degrade the said polymers. For example, different screening methods are utilised for prospecting the biodegradation genes from the metagenome of unique environmental niches. These methods are classified as functional and sequence-based screening (Fig. 6.2).

6.3.1 Sequence-Based Screening

The sequence-based screening method utilises the metagenomic sequencing data obtained through shotgun sequencing and similar methods. The reads obtained from the shotgun sequences can be annotated to identify the functional genes of interest from the metagenome (Kim et al. 2007). Thus, the sequence-based screening mainly consists of two steps: identification of the metagenomic reads with desired sequence (a.k.a. gene prediction) and linking the desired sequences to a database (a.k.a. gene annotation) (Sharpton 2014). Kim et al. (2007) followed the sequence-based metagenomic approach for identifying the P450 monooxygenase from the soil sample from the Keryong Mountains in Korea. They constructed a soil metagenome library in the fosmid backbone of pCC1FOS. They pooled the library clones and extracted the fosmids from each pool, screened the pool using the PCR technique

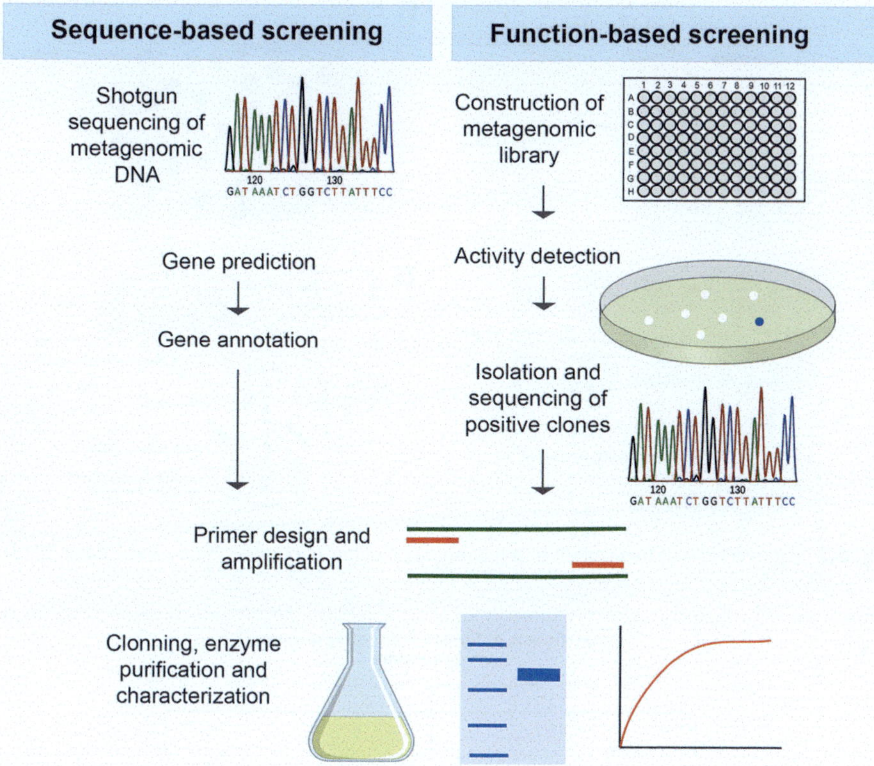

Fig. 6.2 The outline of the metagenome screening for new thermozymes (Courtesy: Decastro et al. 2016)

with CYP-specific primers and cloned the amplicons using the TA cloning method. The group eventually recognised a novel CYP gene, *syk181* from the metagenome. Heterologous expression of the gene *syk181* revealed a noteworthy hydroxylase activity on naphthalene, phenanthrene and various fatty acids.

6.3.2 Function-Based Screening

The function-based metagenomic tool does not rely on the genomic sequence data or sequence-similarity with the previously recognised genes. Instead, this approach depends on the expression profile of the metagenomic library clones. Through this approach, there is a considerable potential to recognise new gene fragments coding for already-identified or novel functions. The functional screening approach comprises of phenotype-based screening. This includes screening by substrate-induced gene expression (SIGEX) (Meier et al. 2016; Pushpanathan et al. 2014), and metabolite expression profiles (METREX) (Jones and Marchesi 2007). Functional studies of metagenomic libraries using agar plates deliver an unpretentious

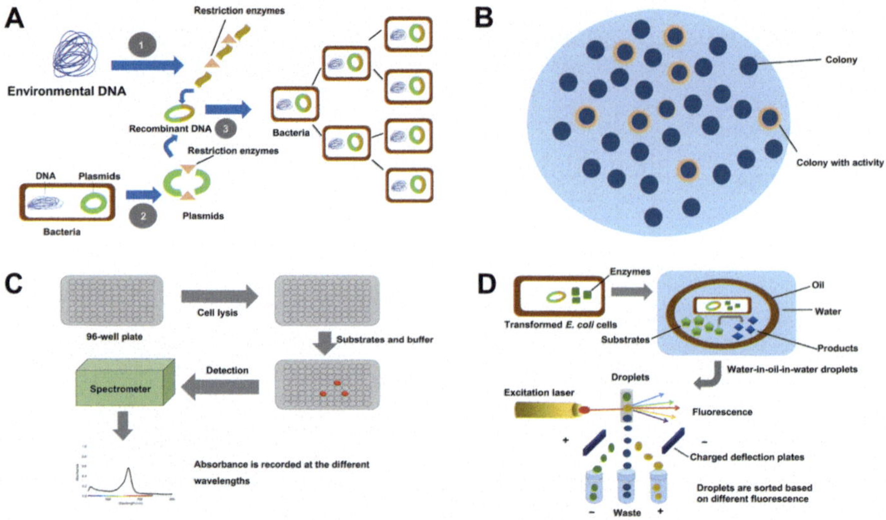

Fig. 6.3 An overview of metagenomic screening approaches (Courtesy: Ngara and Zhang 2018)

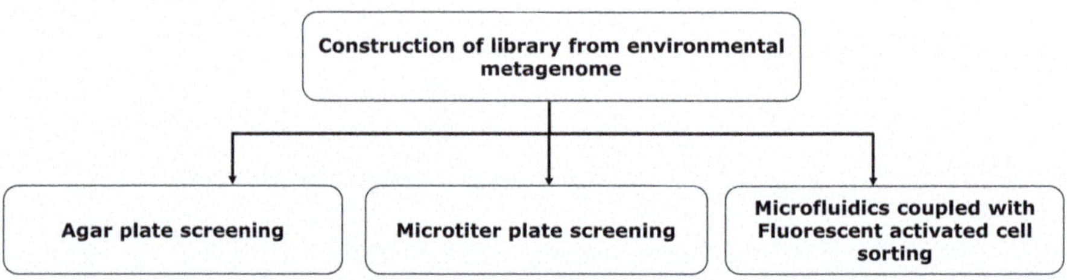

Fig. 6.4 Outlines of commonly used function-based screening approaches

and upfront tactic to find novel enzymes that can be functional in various conditions and on various substrates of interest (Uchiyama and Miyazaki 2009) (Figs. 6.3 (steps summarize in a-d) and 6.4). This approach has enabled the identification of different hydrolytic enzymes for industrial and environmental applications (Table 6.1) (Popovic et al. 2015).

The functional screening approach in metagenomics allows the detection of novel enzymes which are not predicted using the sequence-based approach (Lam et al. 2015). The successful execution of the function-based screening depends upon the size of the genes, the type and the capacity of the vector. Large gene inserts are basically to comprise complete genes with their operons, which allows, the expression of novel enzymes. Along with the success of the functional screening, there are specific challenges encountered during the process like quality and length of gene insert, the proportion of clones to identify whole microbial communities on any given samples, selection and choice of the expression host. Though the frequently

Table 6.1 Function-based screening of metagenomes reported for enzyme production

S. No.	Functional screening	References
1	DNA endonucleases	Mtimka et al. (2020)
2	Extradiol Dioxygenases	Sidhu et al. (2019)
3	β-galactosidase	Cheng et al. (2017)
4	(hemi)cellulases	Maruthamuthu et al. (2016)
5	Periplasmic α-amylase	Pooja et al. (2015)
6	Promiscuous enzymes (hydrolases for sulfate monoesters and phosphotriesters)	Colin et al. (2015)
7	Cold-active enzymes (β-galactosidases, α-amylases and a phosphatase)	Vester et al. (2014)
8	Chitinase	Hjort et al. (2014)
9	Alkaline-stable family IV lipase	Peng et al. (2014)
10	Esterases	Ouyang et al. (2013)
11	Carboxylic ester hydrolases	Biver and Vandenbol (2013)
12	Serine protease (Alkaline)	Biver et al. (2013)
13	B-galactosidase	Wang et al. (2012)
14	Xylanase	Cheng et al. (2012)
15	Lipase (moderately thermostable)	Faoro et al. (2012)
16	Tannase (halotolerant and moderately thermostable)	Yao et al. (2011)
17	Serine protease	Neveu et al. (2011)
18	Dietary fibre catabolic enzymes (beta-glucanase, hemicellulase, galactanase, amylase, or pectinase)	Tasse et al. (2010)

used *E. coli* strains have comfortable necessities for promoter recognition and translation (Uchiyama and Miyazaki 2009).

However, various other tasks hinder the detection of genes for metabolites, such as the weak expression of desired genes and low product yield (Uchiyama and Miyazaki 2009). The conventional agar-plate screening approach has a limitation due to low sensitivity and low-to-medium throughput. For overcoming such challenges, various methods have been adopted, such as fluorescence-activated cell sorting (FACS)-driven screening, and microfluidics-based screening.

6.4 Strategies Allied to Metagenomics

6.4.1 Metatranscriptomics

Metatranscriptomics is the study of mRNA transcripts of a population of microorganisms and the expression of genes in the communities native to different environments (Carvalhais et al. 2012). Metatrascriptomics approaches provide

Fig. 6.5 Workflow of the environmental sample metatranscriptomic analysis

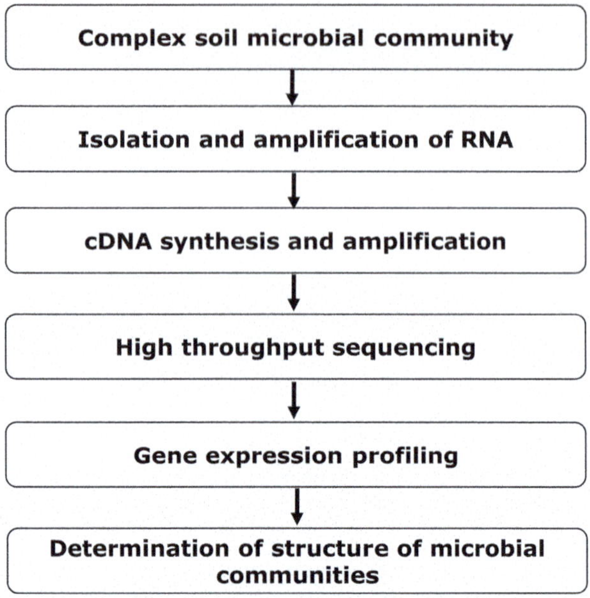

crucial insights into the dynamically expressed genes in a population and therefore, act as appropriate precursors to the functional-based screening under the circumstances. In metatranscriptomics, the RNA is reverse-transcribed into cDNA and then sequenced as a metagenome. In this method, the whole sample presents a catalogue of the gamut of the genes expressing in an environmental sample. Processing the RNA of a given microbial community to study the transcriptome is a sensitive protocol because of the difficulties such as the recovery and handling of mRNA, mRNA stability and yield (Moran 2009). Metatranscriptomics permits in-depth data about the possible expression of genes at the selection time of samples.

Simon and Daniel (2011) demonstrated that metatranscriptomics is affordable and it permits the profiling and quantification of the whole-genome expression of a microbial community. The benefit of metatranscriptomics is that it provides the data of the differences in the active functions of different microbial populations which seem to be identical per the microbial composition. Pyrosequencing has been used and optimised to study the active genes in a microbial community and their functions (Leininger et al. 2006). The importance of small RNAs in various environmental cycles like carbon metabolism and nutrient attainment has been evaluated using metatranscriptomic profiling of the Hawaii Ocean station (Shi et al. (2009). The general workflow (Fig. 6.5) of the metatranscriptomic analysis is as follows:

Metatrascriptomics is helpful to recognize active metabolic pipelines related to the microbial populations in any definite environmental sample. It reflects an additional enlightening view, revealing facts about the communities which are transcriptionally active (Bashiardes et al. 2016).

6.4.2 Metaproteomics Approach

Metaproteomics deals with the study of all protein present in the environmental sample. High-resolution mass spectroscopy is coupled to obtain enzyme digested protein profiling. Various reports have compiled the assessment of the metaproteomics tool to study the metagenome. In-depth knowledge of the proteins present in any sample enables the depiction of diverse developments like degradation, real-time protein encoding by genes, and protein modifications taking place during the bioremediation. Metaproteomics enables the detection and estimation of expressed genes, understanding about working of metabolic pathways utilized in environmental communities as proteins are regarded as crucial mediators of the metabolic pathways (Fig. 6.6) (Wilkins et al. 1996).

Highly inclusive metaproteomics experiments have been planned and showed by Ram et al. (2005), who analysed more than 2000 protein extraction of a native acid mine drainage. However, it is a cumbersome task to detect and recognize all the proteins recovered from a rich environmental microbial population (Simon and Daniel 2011). Various microbial communities, high level of protein expression level, genetic heterogeneity in microbial populations are the key challenges in metaproteomics studies. Despite all these trials, metaproteomics has a vast possibility to connect the genetic multiplicity and actions of microbial populations with their influence on ecosystem balance. Maron et al. (2007) described that even after all these scientific progress, majority of the proteomics studies were carried out under laboratory settings and do not justify the varied interactions going among the microbial populations. Thus, improved methods are essential for the *In situ* identification and assessment of the full range of protein expression at the level of microbial communities.

Fig. 6.6 General workflow outline of the Metaproteomics

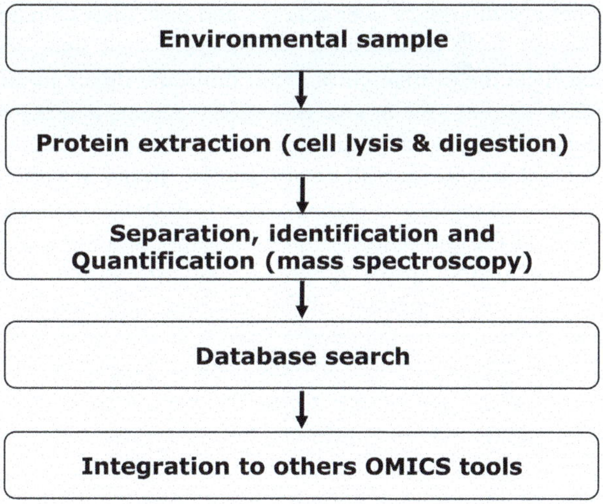

6.4.3 Metabolomics

Metabolomics is the field of study involving the identification and quantification of the *metabolites*, substrates, intermediates and end products of the metabolic pathways (Aguiar-Pulido et al. 2016). Metabolomics is also known as the superglue of all the omics. The metabolome is an appropriate signature of the health and structural composition of a microbial community, which allows the estimation of the microbial activities and functions of the population. Interestingly, metabolomics also provides knowledge about the potential of the microbiome of degradation of the recalcitrant molecules. It is also fascinating that the metabolome can represent signalling processes through complicated communication among bacteria, like quorum sensing, which narrates gene expression behaviour to changes in cell population density.

On the other hand, metabolic sketching can give a prompt information of the physiological state of that cell, and therefore, metabolomics offers a straight "functional information of the physiological state" of any community. In the end, it is the metabolome, which is the fundamental biochemical layer of the genome, transcriptome and proteome, that imitates all the data expressed and controlled by all other omic tools. The metabolome is the adjoining connection to the phenotype of microbial communities.

6.4.4 Fluxomics

Metabolic flux refers to the degree of metabolite conversion in a metabolic pathway. Fluxomics defines the numerous methods that pursue to control the degree of metabolic reactions inside a biological cell. The complete set of metabolic fluxes, or fluxome, characterizes as an active depiction of the cellular phenotype. A major benefit of fluxomics over genome and proteomics studies is that it relies on the data obtained from metabolites, which are distant than genes or proteins. Fluxomics objects at the estimation of small molecule fluxes through metabolic pathways and offers access to the *in vivo* action of reactions and networks in integral living cells. Metabolomic and fluxomic are very current omic claims (Winter and Krömer 2013). Metabolomic analyse all the metabolites found in a microbial community. Fluxomic attempts to regulate metabolic fluxes. Both the tools are recognized as endpoint on behalf of the end outcomes of gene expression, available protein concentration and reaction kinetics, pathway directive and metabolite yield at any given point of time. There is some limitation of the fluxomic apprehensions that is availability and the cost of the tracer compounds. Therefore, fluxomic approach is restricted to limited carbon sources and typically sugar-based processes. Secondly, another limitation is the analytical assessment. Certainly, pattern analysis is finnicky and could lead to imperfect data during understanding of results.

6.5 Microbial Community Profiling of Contaminated Sites

In the last decade, various sequencing tools have been identified as the choice of strategy to study the microbial populations in a contaminated site. Current initiative regarding new sequencing approaches permits us to mine the entire microbial population predominantly living in the polluted sites and has transformed metagenomics (Fig. 6.7; Table 6.2). Pyrosequencing approach is the first choice over traditional Sanger sequencing due to correctness, flexibility, high-end processing, time constraints and operational cost (Droege and Hill 2008). Pyrosequencing methods revealed the microbial population composition from a diesel-contaminated site and found as Proteobacteria, Firmicutes, Actinobacteria, Acidobacteria, and Chloroflexi (Sutton et al. 2013).

6.6 Metagenomics in Bioremediation of Environmental Pollutants

6.6.1 Case Study 1

Lu et al. (2017) reported microbial communities of chemically polluted estuarine sediments of rivers Oujiang and Jiaojiang in East China Sea. The pollutants present in the sediments revealed that microbial community diversity is varied from non-contaminated site to the contaminated sediments. Polycyclic aromatic hydrocarbons (PAHs) and nitrobenzene were undesirably correlated with the bacterial population. The leading population in the sediments was Gamma proteobacteria. The genomic annotation found that the various enzymes profiles were dominant estuarine sediments, which enlarged significantly, like 2-oxoglutarate synthase, acetolactate synthase, inorganic diphosphatase, and aconitate hydratase.

Chemical pollutants of Chinese estuary have a possibility to diminish the natural variability in the microbial population that exist amongst the estuarine sediments.

6.6.2 Case Study 2

The abundance, multiplicity, and dispersal of biodegradation genes in activated sludge from two effluent wastewater treatment units were reported by metagenomic analysis (Fang et al. 2013). The findings showed that the abundance and multiplicity of biodegradation genes in activated sludge diverse with the treatment plant and time of the sampling. The P450 genes was the major abundant genes in the biodegradation process. It was found that 87 bacterial genera possibly proficient of degrading pollutants were typically allied with Proteobacteria (59.8%), Bacteroidetes (17.2%), and Actinobacteria (9.2%). Mycobacterium, belonging to Actinobacteria, was found to be the major abundant genus (23.4%). The present approach could be utilised to monitor an activated sludge potential for organic pollutants and to assess the efficiency of the effluent treatment plant.

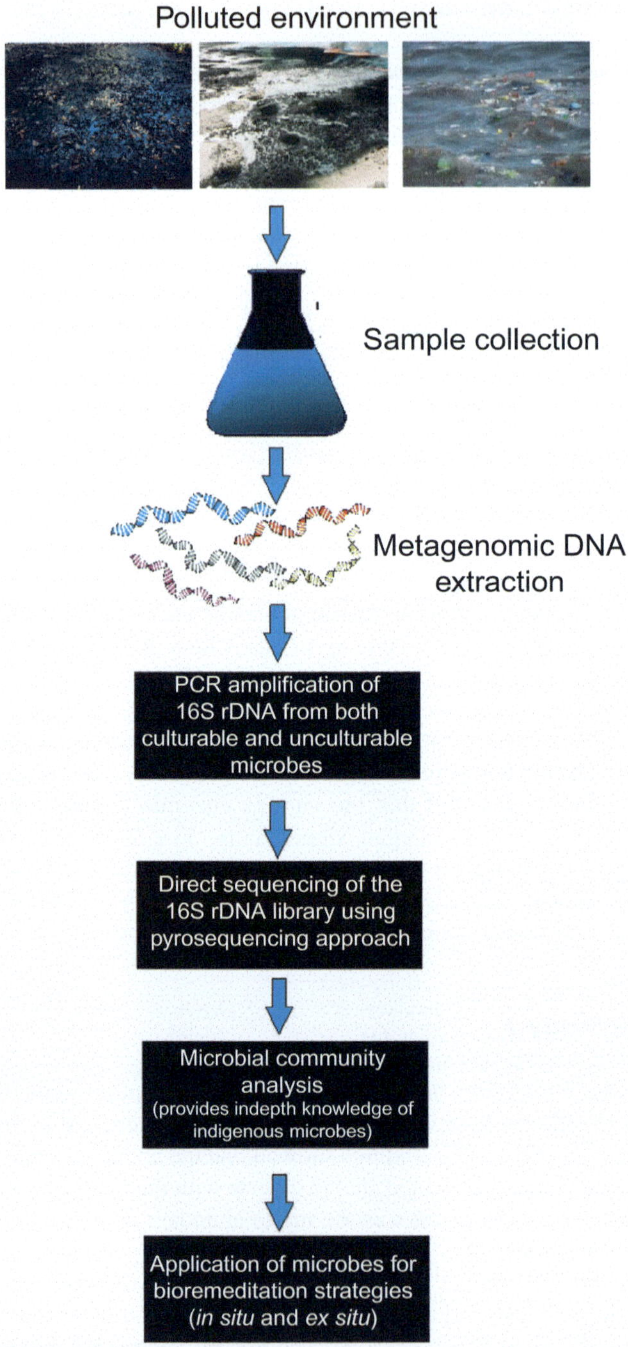

Fig. 6.7 Microbial population analysis of the contaminated sited by sequencing (Courtesy: Pushpanathan et al. 2014)

Table 6.2 Various reports of metagenomic studies about bioremediation genes

S. No.	Bioremediation genes	Samplin sites	References
1	Metallothionein genes	Soil microbiome	Li et al. (2020)
2	long-chain hydrocarbon bioremediation (metagenome-assembled genomes of *Marinobacter adhaerens* t76_800)	Marine samples	Lopes et al. (2020)
3	Malachite green degradation genes	Mangrove sediments	Qu et al. (2018)
4	Carbamate degrading enzymes	Bovine rumen microbiome	Ufarté et al. (2017)
5	Cellulase (mining bioremediation system)	Biochemical reactor	Mewis et al. (2013)
6	Pyrethroid-hydrolyzing enzyme (thermostable)	River Basin	Fan et al. (2012)
7	Hydrocarbon-degrading genes	Diesel contaminated sites	Yergeau et al. (2012)
8	Extradiol dioxygenase	Activated sludge	Suenaga et al. (2009)

The precedence pollutant biodegradation ability of microbial communities with potential biodegradation genes in activated sludge are typically revealed using conventional cultivation-dependent methods, traditional PCR (Felföldi et al. 2010). Microbial bioremediation includes various processes such as dehalogenation, dealkylation, hydrolysis, oxidation, reduction, ring cleavage, conjugation, and methylation (Jechalke et al. 2011).

6.6.3 Case Study 3

Jadeja et al. (2019) performed metagenomic analysis for significant analysis of microbial population from effluent treatment plant. Activated sludge process is a kind of ecosystem which has a pool of biodegrading communities. They have analysed such ecosystem a treatment plant where more than 200 industries waste effluent. The screening resulted in predicting approximately 30 degradative pathways in wastewater. Results were stretched to plan a bioremediation approach using 4-methylphenol, 2-chlorobenzoate, and 4-chlorobenzoate as target molecules. The experiments highlight the significance to determine the microbial communities activated sludge to yoke its hidden potential. Metagenomics tools permits an indistinct understanding into the complex degrading pathways functioning at the wastewater treatment site and the comprehensive certification of genes permits the activated inoculum to be explored as a bioresource.

6.7 Conclusion

Microorganisms theatres a vital role in the working of the ecosystem. Bioremediation approaches exploit the metabolic activity of microbial cells to clean up environmental polluted sites. The conventional approach of pollutant removal includes the isolation and screening of indigenous organisms proficient of degrading pollutants. But the traditional culture dependent methods are only able to culture, 1% of microbes in a contaminated sample, and thus, a huge amount of microbial communities in the environment confront cultivation in axenic culture. An inclusive knowledge of microbial community composition and its functional is essential to attain an effective and consistent approach to clean up environmental pollutants from polluted sample. Metagenomic tools, composed with developing high-throughput sequencing knowledges have elaborated understanding into the complex exchanges among diverse microbial populations and their metabolic ability, which could be utilized in different bioremediation approaches for effective exclusion contaminants from the environment.

References

Aguiar-Pulido V, Huang W, Suarez-Ulloa V et al (2016) Metagenomics, metatranscriptomics, and metabolomics approaches for microbiome analysis: supplementary issue: bioinformatics methods and applications for big metagenomics data. Evol Bioinforma 12:EBO-S36436

Antizar-Ladislao B (2010) Bioremediation: working with bacteria. Elements 6:389–394

Bashiardes S, Zilberman-Schapira G, Elinav E (2016) Use of metatranscriptomics in microbiome research. Bioinform Biol Insights 10:BBI-S34610

Biver S, Vandenbol M (2013) Characterization of three new carboxylic ester hydrolases isolated by functional screening of a forest soil metagenomic library. J Ind Microbiol Biotechnol 40:191–200

Biver S, Portetelle D, Vandenbol M (2013) Characterization of a new oxidant-stable serine protease isolated by functional metagenomics. Springerplus 2:410

Carvalhais LC, Dennis PG, Tyson GW, Schenk PM (2012) Application of metatranscriptomics to soil environments. J Microbiol Methods 91:246–251

Cheng F, Sheng J, Dong R et al (2012) Novel xylanase from a holstein cattle rumen metagenomic library and its application in xylooligosaccharide and ferulic acid production from wheat straw. J Agric Food Chem 60:12516–12524

Cheng J, Romantsov T, Engel K et al (2017) Functional metagenomics reveals novel β-galactosidases not predictable from gene sequences. PLoS One 12:e0172545

Colin P-Y, Kintses B, Gielen F et al (2015) Ultrahigh-throughput discovery of promiscuous enzymes by picodroplet functional metagenomics. Nat Commun 6:1–12

DeCastro M-E, Rodríguez-Belmonte E, González-Siso M-I (2016) Metagenomics of thermophiles with a focus on discovery of novel thermozymes. Front Microbiol 7:1521

Droege M, Hill B (2008) The Genome Sequencer FLXTM System—Longer reads, more applications, straight forward bioinformatics and more complete data sets. J Biotechnol 136:3–10

Fan X, Liu X, Huang R, Liu Y (2012) Identification and characterization of a novel thermostable pyrethroid-hydrolyzing enzyme isolated through metagenomic approach. Microb Cell Factories 11:33

Fang H, Cai L, Yu Y, Zhang T (2013) Metagenomic analysis reveals the prevalence of biodegradation genes for organic pollutants in activated sludge. Bioresour Technol 129:209–218

Faoro H, Glogauer A, Couto GH et al (2012) Characterization of a new Acidobacteria-derived moderately thermostable lipase from a Brazilian Atlantic Forest soil metagenome. FEMS Microbiol Ecol 81:386–394

Felföldi T, Székely AJ, Gorál R et al (2010) Polyphasic bacterial community analysis of an aerobic activated sludge removing phenols and thiocyanate from coke plant effluent. Bioresour Technol 101:3406–3414

Gadd GM (2010) Metals, minerals and microbes: geomicrobiology and bioremediation. Microbiology 156:609–643

Handelsman J (2004) Metagenomics: application of genomics to uncultured microorganisms. Microbiol Mol Biol Rev 68:669–685

Handelsman J, Rondon MR, Brady SF et al (1998) Molecular biological access to the chemistry of unknown soil microbes: a new frontier for natural products. Chem Biol 5:R245–R249

Hjort K, Presti I, Elväng A et al (2014) Bacterial chitinase with phytopathogen control capacity from suppressive soil revealed by functional metagenomics. Appl Microbiol Biotechnol 98:2819–2828

Jadeja NB, Purohit HJ, Kapley A (2019) Decoding microbial community intelligence through metagenomics for efficient wastewater treatment. Funct Integr Genomics 19:839–851

Jechalke S, Rosell M, Martínez-Lavanchy PM et al (2011) Linking low-level stable isotope fractionation to expression of the cytochrome P450 monooxygenase-encoding ethB gene for elucidation of methyl tert-butyl ether biodegradation in aerated treatment pond systems. Appl Environ Microbiol 77:1086–1096

Jones BV, Marchesi JR (2007) Accessing the mobile metagenome of the human gut microbiota. Mol BioSyst 3:749–758

Karigar CS, Rao SS (2011) Role of microbial enzymes in the bioremediation of pollutants: a review. Enzyme Res 2011:805187

Kim BS, Kim SY, Park J et al (2007) Sequence-based screening for self-sufficient P450 monooxygenase from a metagenome library. J Appl Microbiol 102:1392–1400

Kumar P, Sharma PK, Sharma PK, Sharma D (2015) Micro-algal lipids: a potential source of biodiesel. J Sustain Bioenergy Syst 2:135–143

Lam KN, Cheng J, Engel K et al (2015) Current and future resources for functional metagenomics. Front Microbiol 6:1196

Leininger S, Urich T, Schloter M et al (2006) Archaea predominate among ammonia-oxidizing prokaryotes in soils. Nature 442:806–809

Leung M (2004) Bioremediation: techniques for cleaning up a mess. BioTeach J 2:18–22

Li X, Islam MM, Chen L et al (2020) Metagenomics-guided discovery of potential bacterial metallothionein genes conferring Cu/Cd resistance from soil microbiome. Appl Environ Microbiol 86(9):e02907

Lopes EM, Fernandes CC, de Macedo Lemos EG, Kishi LT (2020) Reconstruction and in silico analysis of new Marinobacter adhaerens t76_800 with potential for long-chain hydrocarbon bioremediation associated with marine environmental lipases. Mar Genomics 49:100685

Lorenz P, Eck J (2005) Metagenomics and industrial applications. Nat Rev Microbiol 3:510–516

Lovley DR (2003) Cleaning up with genomics: applying molecular biology to bioremediation. Nat Rev Microbiol 1:35–44

Lu X-M, Chen C, Zheng T-L (2017) Metagenomic insights into effects of chemical pollutants on microbial community composition and function in estuarine sediments receiving polluted river water. Microb Ecol 73:791–800

Maron P-A, Ranjard L, Mougel C, Lemanceau P (2007) Metaproteomics: a new approach for studying functional microbial ecology. Microb Ecol 53:486–493

Maruthamuthu M, Jiménez DJ, Stevens P, van Elsas JD (2016) A multi-substrate approach for functional metagenomics-based screening for (hemi) cellulases in two wheat straw-degrading microbial consortia unveils novel thermoalkaliphilic enzymes. BMC Genomics 17:86

Meier MJ, Paterson ES, Lambert IB (2016) Use of substrate-induced gene expression in metagenomic analysis of an aromatic hydrocarbon-contaminated soil. Appl Environ Microbiol 82:897–909

Mewis K, Armstrong Z, Song YC et al (2013) Biomining active cellulases from a mining bioremediation system. J Biotechnol 167:462–471

Moran MA (2009) Metatranscriptomics: eavesdropping on complex microbial communities. Microbe 4:7

Mtimka S, Pillay P, Rashamuse K et al (2020) Functional screening of a soil metagenome for DNA endonucleases by acquired resistance to bacteriophage infection. Mol Biol Rep 47:353–361

Neveu J, Regeard C, DuBow MS (2011) Isolation and characterization of two serine proteases from metagenomic libraries of the Gobi and Death Valley deserts. Appl Microbiol Biotechnol 91:635–644

Ngara TR, Zhang H (2018) Recent advances in function-based metagenomic screening. Genomics Proteomics Bioinformatics 16:405–415

Ouyang L-M, Liu J-Y, Qiao M, Xu J-H (2013) Isolation and biochemical characterization of two novel metagenome-derived esterases. Appl Biochem Biotechnol 169:15–28

Peng Q, Wang X, Shang M et al (2014) Isolation of a novel alkaline-stable lipase from a metagenomic library and its specific application for milkfat flavor production. Microb Cell Factories 13:1

Pooja S, Pushpanathan M, Jayashree S et al (2015) Identification of periplasmic α-amlyase from cow dung metagenome by product induced gene expression profiling (pigex). Indian J Microbiol 55:57–65

Popovic A, Tchigvintsev A, Tran H et al (2015) Metagenomics as a tool for enzyme discovery: hydrolytic enzymes from marine-related metagenomes. In: Krogan NJ, Babu M (eds) Prokaryotic systems biology. Springer, Cham, pp 1–20

Pushpanathan M, Jayashree S, Gunasekaran P, Rajendhran J (2014) Microbial bioremediation: a metagenomic approach. In: Das S (ed) Microbial biodegradation and bioremediation. Elsevier, Amsterdam, pp 407–419

Qu W, Liu T, Wang D et al (2018) Metagenomics-based discovery of malachite green-degradation gene families and enzymes from mangrove sediment. Front Microbiol 9:2187

Ram RJ, VerBerkmoes NC, Thelen MP et al (2005) Community proteomics of a natural microbial biofilm. Science 308:1915–1920

Riesenfeld CS, Schloss PD, Handelsman J (2004) Metagenomics: genomic analysis of microbial communities. Annu Rev Genet 38:525–552

Saharan VK, Badve MP, Pandit AB (2011) Degradation of Reactive Red 120 dye using hydrodynamic cavitation. Chem Eng J 178:100–107

Sharma D (2016) Biosurfactants in food. Springer, Cham

Sharma D, Dhanjal DS (2016) Bio-nanotechnology for active food packaging. J Appl Pharm Sci 6:220–226

Sharma D, Dhanjal DS, Mittal B (2017) Development of edible biofilm containing cinnamon to control food-borne pathogen. J Appl Pharm Sci 7:160–164

Sharpton TJ (2014) An introduction to the analysis of shotgun metagenomic data. Front Plant Sci 5:209

Shi Y, Tyson GW, DeLong EF (2009) Metatranscriptomics reveals unique microbial small RNAs in the ocean's water column. Nature 459:266–269

Sidhu C, Solanki V, Pinnaka AK, Thakur KG (2019) Structure elucidation and biochemical characterization of environmentally relevant novel extradiol dioxygenases discovered by a functional metagenomics approach. mSystems 4. https://doi.org/10.1128/mSystems.00316-19

Simon C, Daniel R (2011) Metagenomic analyses: past and future trends. Appl Environ Microbiol 77:1153–1161

Suenaga H, Koyama Y, Miyakoshi M et al (2009) Novel organization of aromatic degradation pathway genes in a microbial community as revealed by metagenomic analysis. ISME J 3:1335–1348

Sutton NB, Maphosa F, Morillo JA et al (2013) Impact of long-term diesel contamination on soil microbial community structure. Appl Environ Microbiol 79:619–630
Tasse L, Bercovici J, Pizzut-Serin S et al (2010) Functional metagenomics to mine the human gut microbiome for dietary fiber catabolic enzymes. Genome Res 20:1605–1612
Techtmann SM, Hazen TC (2016) Metagenomic applications in environmental monitoring and bioremediation. J Ind Microbiol Biotechnol 43:1345–1354
Uchiyama T, Miyazaki K (2009) Functional metagenomics for enzyme discovery: challenges to efficient screening. Curr Opin Biotechnol 20:616–622
Ufarté L, Laville E, Duquesne S et al (2017) Discovery of carbamate degrading enzymes by functional metagenomics. PLoS One 12:e0189201
Venter JC, Remington K, Heidelberg JF et al (2004) Environmental genome shotgun sequencing of the Sargasso Sea. Science 304:66–74
Vester JK, Glaring MA, Stougaard P (2014) Discovery of novel enzymes with industrial potential from a cold and alkaline environment by a combination of functional metagenomics and culturing. Microb Cell Factories 13:72
Vidali M (2001) Bioremediation. an overview. Pure Appl Chem 73:1163–1172
Wang K, Lu Y, Liang WQ et al (2012) Enzymatic synthesis of Galacto-oligosaccharides in an organic–aqueous biphasic system by a novel β-Galactosidase from a metagenomic library. J Agric Food Chem 60:3940–3946
Wilkins MR, Pasquali C, Appel RD et al (1996) From proteins to proteomes: large scale protein identification by two-dimensional electrophoresis and arnino acid analysis. Biotechnology 14:61–65
Winter G, Krömer JO (2013) Fluxomics–connecting 'omics analysis and phenotypes. Environ Microbiol 15:1901–1916
Yao J, Fan XJ, Lu Y, Liu YH (2011) Isolation and characterization of a novel tannase from a metagenomic library. J Agric Food Chem 59:3812–3818
Yergeau E, Sanschagrin S, Beaumier D, Greer CW (2012) Metagenomic analysis of the bioremediation of diesel-contaminated Canadian high arctic soils. PLoS One 7:e30058

Metagenomics and Enzymes: The Novelty Perspective

7

Daljeet Singh Dhanjal, Reena Singh Chopra, and Chirag Chopra

Abstract

Microbial enzymes have become an integral part of the industrial processes and are significantly influencing our life. Exploration of microbial enzymes has seen various improvements over the past years like directed evolution, omics studies, recombinant DNA technology and metagenomics. The major challenge with microbial enzyme is that only 1% of the microbial community has been explored for being culturable, whereas, rest of 99% still remains mystery and demands for exploration. The metagenomic approach has emerged as valuable tool to untapped the hidden microbial community. This process involves the direct isolation of the DNA from various niches, followed by construction of metagenomic library and extraction of the information via sequence and functional based approaches. Numerous novel biocatalyst and metabolites have been obtained through this approach. Hence, mining of the microbial genomes can expand our horizon to find industrially important enzymes. This chapter intends to provide the comprehensive information about metagenomics and its potential exclusively in obtaining the novel microbial enzymes.

Keywords

Bioprospecting · Bioinformatics · Enzymes · Metagenomics · Microbial diversity

D. S. Dhanjal · R. S. Chopra · C. Chopra (✉)
School of Bioengineering and Biosciences, Lovely Professional University, Phagwara, Punjab, India
e-mail: reena.19408@lpu.co.in; chirag.18298@lpu.co.in

R. S. Chopra et al. (eds.), *Metagenomics: Techniques, Applications, Challenges and Opportunities*, https://doi.org/10.1007/978-981-15-6529-8_7

7.1 Introduction

Presently, microbial enzymes are used in several industries ranging from chemical, food and pharmaceutical industries to manufacture products like detergent, food, textile, paper. In 2015, industrial enzymes accorded the net profit of five billion USD and is expected to reach the net worth of 10.7 billion USD by 2024, because of the increasing demand of novel enzymes with effective traits and performance (IEA 2020). As these enzyme-based processes reduces the use of toxic compounds and harsh conditions and reduce the overall cost of process. Additionally, it also increases the efficiency and improve the product recovery process (Singh et al. 2016). Microbes are known to account for a significant portion of the biomass on this planet, and it has been estimated that about 4–6 $\times$ 10^{10} prokaryotes cells inhabitants in different niches like Antarctic ice, brackish water, hot springs and hydrothermal vents. On the basis of environmental DNA (eDNA) analysis, it has been estimated that only 0.1–1% of prokaryotes are culturable through conventional microbial approaches (Dhanjal et al. 2017; Anand et al. 2019). Even, some of the eukaryotic microbes are unculturable, the reason for which is still unknown (McNichol et al. 2018).

In 1998, the introduction of metagenomics and its allied fields like metaproteomics and metatranscriptomics have paved the way to untapped the unculturable microbial community and obtain the novel enzyme (Simon and Daniel 2011). Metagenomics has now emerged as the revolutionary technology to discover new enzymes and metabolites form eDNA. This approach involves the direct isolation of eDNA from the sample and provide the complete genetic information of isolated microbial genome (Dhanjal and Sharma 2018). Till now, various diverse and unique environments have been explored and new niches are being explored for unveiling the new enzymes or metabolites (Handelsman 2004). Therefore, metagenomics could serve as the effective method for untapping the unexplored microbial community and find the new enzymes with improved catalytic activity (Sarangi et al. 2019). Moreover, the advancements in high-throughput sequencing techniques also help in precise identification of genes coding for the enzyme or metabolites from the complex metagenome (Ferrer et al. 2008). This chapter aims to provide the brief overview of metagenomic approach and discuss about its potential in discovering novel enzyme. Additionally, it will also provide the information about the different enzymes obtained by the metagenomic approach. The limitations of metagenomic approach and their solutions will also be discussed.

7.2 The Dilemma Between Known, Unknown and Engineered Enzymes

To inspect the enzymes of interest for suitable chemical reactions, the first approach is to investigate the already available commercial enzymes (Gurung et al. 2013). These enzymes are mainly derived from biological sources among which the diverse microorganisms account for 88% of the industrial enzyme production while the remaining are extracted from plants (4%) and animals (8%) (Raveendran et al. 2018; Mukherjee et al. 2018). But the absence of diversified enzymes is a major limitation. However, recent advancements in various fields like microbiology, biochemistry and molecular biology have widened the horizons for the development of more efficient enzymes (Singh et al. 2016).

The second approach involves the exploitation of previously available enzymes to get the desired characteristics (Robinson 2015). Previous experimental and theoretical studies in the literature suggest that novel enzymes can be designed that naturally does not exist in nature or currently available enzymes can be altered to provide them with characteristics differing from the native enzyme (Maria-Solano et al. 2018). Thus, by designing enzymes, it is possible to increase their specificity towards the substrate and enhance the enzyme activity. By employing recombinant techniques, modifications can be done at the molecular level. Such alterations can result in structural changes such as a modified active site with more specificity towards substrate, thereby, yielding a novel enzyme with improved characteristics (Baweja et al. 2016). The process of protein engineering is widely used to produce new or better enzymes. Protein engineering involves production of mutant DNA libraries, in which, the most suitable enzyme-producing clone is selected based on desired requirements. When this technique is combined with high-throughput screening, a technologically advanced system, it can offer the advantage of limited risk and easy production of the mutant library (Wójcik et al. 2015). Moreover, this method has a high success rate of selection of the desired mutant from the mutant library. While the experimental studies suggest the production of novel enzymes, but it is highly unlikely that the engineered enzymes will be introduced in industries (Valetti and Gilardi 2013). Moreover, the use of protein engineering for all available enzymes will be cost-intensive, further limiting its industrial applicability (Woodley 2013).

A third approach involves the bioprospecting of new enzymes. Unlike protein engineering, this approach allows for exploring novel proteins. Given the presence of diverse enzymes in nature, efficient methods need to be developed to explore suitable enzymes for a range of processes (Engqvist and Rabe 2019). Microbial diversity is believed to be the most significant source of biological products, especially enzymes. Indeed, the constant evolution of ecosystem life gives rise to a great biodiversity (National Research Council (US) Committee on Noneconomic and Economic Value of Biodiversity 1999). However, the actual scope of biodiversity of microorganisms is still unknown, revealing the classification of only approximately 11,000 Archaeal and bacterial species so far and, on an average, around 600 new species are reported annually. On this rate, researchers predict that the classification of 30 M microbial species would require a period of 1000 years while the rest 25% species reside in large water bodies like seas and oceans (Louca et al. 2019). These numbers indicate the unexplored microbial potential in terms of their activity and news genes. Despite this, researchers have recognized the immense hidden potential of microbial population and role they play in improving the economy in future.

7.3 Metagenomics and Its Approaches

Jo Handelsman was the first scientist to introduce the term metagenomics to the world. With the development of metagenomics, came the realization of the immense hidden potential of diverse microbial communities inhabiting a variety of environmental niches (Escobar-Zepeda et al. 2015). Metagenomics is primarily concerned with the complete genetic composition of diverse microorganisms present in the environment. It emphasizes on the extraction of DNA of whole community and detailed genetic analysis of the obtained data (Kunin et al. 2008). The extensive research in field of metagenomics has led to comprehensively understand the microbial diversity inhabiting a particular habitat. It has also aided in discovering novel or new enzymes and further enhanced the knowledge of microbial life (Malla et al. 2019). Projects of metagenomic sequencing have exponentially increased with the advent of cost effective and high-throughput screening approaches such as the project concerning metagenomic comparison of 42 viromes and 45 different microbiomes, Sorcerer II Global Ocean Sampling etc. (Rosen et al. 2009). Since the discovery and establishment of effective metagenomic methods, they are being widely employed to explore new enzymes and consolidation of these methods with recent sequencing technologies will soon replace the conventional methods, especially culture dependent ones. The metagenomic methods hold great potential to provide wide volume of information and data (Zhou et al. 2015).

To organise the numerous sequences obtained from high-throughput metagenomic projects, computational tools and bioinformatic methods are widely used. Bioinformatic methods also assist in establishing phylogenetic relationship among different sequences (Oulas et al. 2015). In case of a discovery of a new sequence, biotechnological techniques are employed for detailed analysis such as characterization of genes and its products, cloning etc. There has been an exponential rise in the discovery of genes using this integrates metagenomic approach (Alberts et al. 2002). The isolation of novel enzymes has been done from diverse metagenomes such as insects (lignocellulolytic enzymes), aquatic environment, gut of vertebrates, various soils, extreme environments (acid mine drainage, halo-alkaline waterbodies) etc. (Lee and Lee 2013).

Based on the target, there are two basic metagenomic approaches, homology and functional screening. Both approaches require development of a library, in which cloning of complete metagenomic DNA is done. Subsequently, it is screened for new genes with either new or known properties including its encoded products like antibiotics (Ab), Ab-resistant genes, lipases, oxidoreductases etc. (Ngara and Zhang 2018).

7.3.1 Sequence-Based Screening

Colony hybridization and PCR are some of the very basic techniques utilized to identify the novel enzymes via homology sequencing. For identification, the knowledge about the genomic sequence is a prerequisite as it is required for designing the primer (Kotik 2009). Therefore, this method is only efficient in identifying the mutants of the known enzymes. This method has been utilized to identify the genes of numerous novel enzymes like dioxygenases, glycerol dehydratases, hydrazine oxidoreductases, nitrite reductases, chitinases, herbicide-degrading genes, proteases, copper resistance enzymes etc. However, technologically advanced sequencing system, high-throughput screen has gradually replaced the conventional hybridization techniques and PCR (Morimoto and Fujii 2009). Using bioinformatic methods, it becomes easy to process the obtained sequence data and can further assist in deducing additional information like enzyme sequence motifs. In comparison to PCR, the high-throughput screening offers more flexibility in homology searches. In case of synthetic metagenomics, the desired gene can be optimized for heterologous expression by functional analysis of existing metagenomic data (Garza and Dutilh 2015).

Novel hydrolases have been discovered from hot environments using sequence-based method. Around 100 potentially novel hydrolases have been identified from metagenomes of 15 hot springs using a combination of functional and sequence-based screening method. Among these isolates, structural and functional characterization of many hydrolases has been done which are enol lactonases, epoxide hydrolases, thermostable carboxylesterases, gluconolactonases, quorum sensing lactonases, and cellulases (Wohlgemuth et al. 2018). Using this approach, thermostable carbonic anhydrase has been discovered from hydrothermal vents. Moreover, a novel acetyl xylan esterase has been isolated from metagenomic DNA of hot desert hypolith using sequencing-based method and using bioinformatic tools, it has been annotated for acetyl xylan esterases (AcXEs) (Yadav et al. 2019).

7.3.2 Function-Based Screening

Functional screening is the most common techniques used to explore new enzymes. The screening is based on the metabolic activities of the clones present in metagenomic library. This method does not require the knowledge about sequence of organisms, therefore allowing easy identification of novel enzymes. This technique employs three different strategies: (a) phenotypic detection of biomolecules (b) heterologous complementation of mutants (c) induced gene expression (DeCastro et al. 2016).

For the phenotypic analysis, libraries of variants are plated on suitable substrate upon which enzyme can act and detection is done by formation of halo or change in color of substrate. to. The clones giving positive results are then identified (Markel et al. 2020). Popovic et al. used this method to obtain clones having activity for carboxylesterase and identified 714 clones. Among these, only 80 clones belonging to 17 distinct families showed esterase activity. Expression of three metagenomic enzymes was observed, whereas seven proteins were found to be positive exhibiting polyester depolymerization activity against polycaprolactone andpolylactic acid (Popovic et al. 2017). Moreover, some studies have reported enzymes from metagenomic DNA library and soil DNA of Arizona that are active against several bacteria. *Ralstonia metallidurans* was used to prepare a library of approximately 700,000 cosmid clones which were screened via inhibition of growth using overlay method. The positive colonies were further investigated for the production certain enzymes like peptidase, lipase, cellulase and glycolytic activities (Iqbal et al. 2014).

Another approach in functional screening is heterologous complementation. The host strain or mutant strains synthesise some components essential for growth and this fact is exploited in heterologous complementation in which the particular gene coding essential components is targeted. This technique offers screening of numerous metagenomic libraries in relatively less time. It is highly sensitive and accurate, and does not produce false results (Vester et al. 2015). This technique has been used to identify novel β-galactosidases by making clones of the metagenomic DNA in a host organism. IncP cosmid was used to build a soil metagenomic library for heterologous complementation to detect β-galactosidase activity in both *E. coli* (Gammaproteobacteria) and *Sinorhizobium meliloti* (Alphaproteobacteria). One β-galactosidase was selected that belonged to glycoside hydrolase family 2 (Cheng et al. 2017).

In 2005, Uchiyama et al. presented another strategy of functional screening based on substrate-induced gene expression screening (SIGEX). SIGEX is a high-throughput screening method that use a combination of which uses fluorescence induced cell sorting and GFP expression system. SIGEX uses the underlying principle that in the presence of specific substrates, the catabolic gene expression is induced and is believed to be regulated by the elements in proximity to catabolic genes (Uchiyama and Watanabe 2008). In SIGEX-based screening, the expression of GFP gene is influenced by cloning the metagenomic DNA in the upstream of GFP gene in the vector. Upon incorporation of substrate, the GFP expressed is induced and the clones showing positive results are selected using fluorescence activated cell sorting. However, involvement of other transcriptional regulators in the process limit the efficiency of this method (Meier et al. 2016).

In 2005, Williamson et al. described a similar technique, metabolite-regulated expression (METREX). In this technique, along with the placement of metagenomic DNA in the vector, a quorum sensing-based biosensor capable of controlling the expression of GFP gene is also inserted which aids in identification of metagenomic clones that produce small molecules. A green fluorescent protein (GFP) is synthesized when metagenomic DNA is expressed beyond threshold value which is then detected using fluorescence microscope (Williamson et al. 2005).

Product-induced gene expression (PIGEX) is another approach that involves the detection of enzymatic activities via GFP expression which is further triggered by formation of a particular product. PIGEX has been used to derive metagenomic DNA from activated sludge utilizing a benzoate-responsive transcriptional regulator (BenR sensor) and used for mining of amidases (Uchiyama and Miyazaki 2010). Using BenR sensor, *E. coli* strains along with 96,000 metagenomic clones were cultured using media enriched with benzamide. When metagenomic clones synthesise benzoate, it reacts with the biosensor present inside the host and a fluorescence is emitted. These advanced techniques have been a great success in identifying the new enzymes with the help of functional based screening approach (Van Der Helm et al. 2018).

7.3.3 Workflow of Metagenomic Approach

In general, most of the enzymes used for industrial processes are obtained from microbes, hence their isolation and identification are the key steps to improving their utilization in different industrial activities. Due to the advent of metagenomics, it has now become feasible to isolate the eDNA to explore the diverse microbial genomic pool, and has unveil the new avenues to find novel enzyme with promising applications in various industrial processes (Coughlan et al. 2015). The steps followed for exploration of the new enzymes from the metagenome are explained in this section and is illustrated in Fig. 7.1.

7.3.3.1 Pre-Treatment of an Environmental Sample

Prior to the initial screening, pre-treatment of the sample is done so that desired clones can be enriched or selective clones could be eliminated. This enrichment approach enhances the specificity of metagenomic DNA from the sample and serve as effective approach to improve the sequence-based screening for exploring novel gene (Gu et al. 2019). However, this approach results in the exclusion of major section of microbial community, as gene of interest involves the small portion of the genetic material during the metagenomic screening process. Thus, enrichment approach could be employed for isolation of the microbial strain, for this substrate

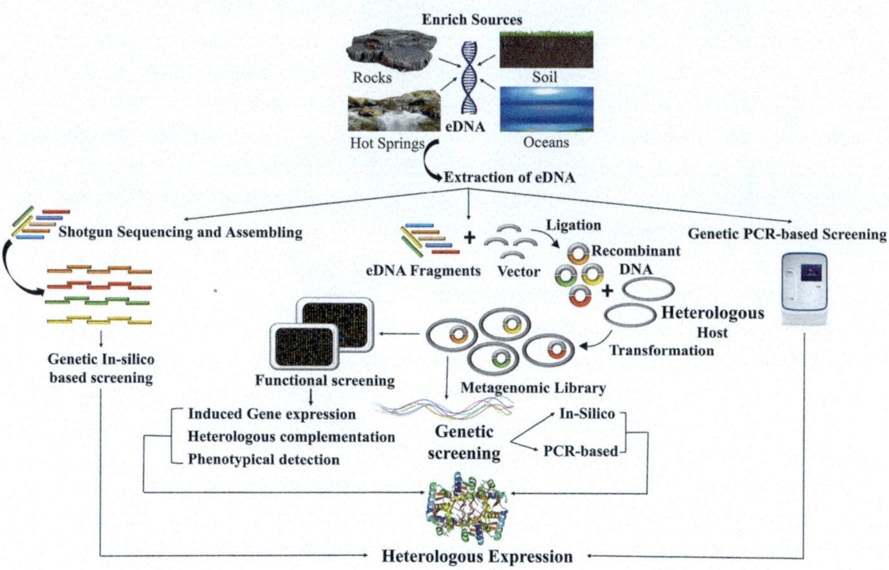

Fig. 7.1 Diagrammatic illustration of the workflow of metagenomic approach

utilization is commonly used (Mukherjee et al. 2017). There are many other methods like differential expression analysis, stable isotope probing (SIP), microarrays, and suppression subtractive hybridization, that are used for enrichment screening. Usually, non-enrichment approach is favoured in experimental studies as they assess the diversity of microbial population (Uhlik et al. 2013).

7.3.3.2 Metagenomic DNA Isolation

There are two main methods for extracting DNA i.e. direct and indirect extraction from the environment. Direct extraction involves the direct lysis of the sample with the help of molecular reagents without culturing the microbes. Whereas, indirect extraction involves physical separation of cells prior to the lysis of the sample for DNA extraction (Roopnarain et al. 2017). Direct extraction procedure is ideally not best option to construct metagenomic library due to poor yield of DNA and inappropriate size (i.e. 1–50 kb) of the DNA stretch because of the destructive nature of molecular reagents (Shamim et al. 2017). In spite of these disadvantages, direct extraction procedure is predominantly used having recovery rate of 15–100 folds better than indirect extraction procedure. Hence, it is important to select the DNA isolation procedure by considering the operational barriers and without compromising the purity as well as integrity of the metagenomic DNA (Gabor et al. 2003).

7.3.3.3 Construction of Metagenomic Library

Construction of metagenomic library predominantly depends upon appropriate host and vector. For the selection of vector, the key feature like copy number, insert capacity, choice of host and the screening procedure are taken into consideration (Lam et al. 2015). Vectors like bacterial artificial chromosomes (BAC), cosmids, plasmids and phagemids are used for constructing metagenomic library. Out of all, plasmid is the most commonly used vector system for insertion and amplification of 15 kb DNA insert. But, in case, when there is need for the amplification of the large stretch of DNA then fosmid (35–45 kb inserts capacity) and, BACs (200 kb inserts capacity) are used. Further, selection of a host is of utmost important for effective cloning and expression of inserted gene (Bajpai 2014). Presently, *Escherichia coli* is the ideal host for constructing metagenomic library, other than that *Pseudomonas* and *Streptomyces* sp. are also used (Kimura 2014).

7.3.3.4 Screening of the Metagenomics Library

On comparing the screening approaches with construction of metagenomic library, it is found to be complex and technically demanding. The frequently used method for screening the metagenomic library are function-based screening, sequence-based screening, Substrate-induced gene-expression screening (SIGEX) and compound configuration screening (Simon and Daniel 2011).

The function-based screening involves the selection of the enzyme based on the reaction catalyzed by them and by assessing the variation in the genetic makeup in contrast to the function of the known enzymes that have been well comprehended (Ahmed et al. 2016). Although, this approach allows us to secure the entire gene/gene cluster coding for the desired trait. But there are certain limitations are also associated with it like transcription, translation and expression of the desired gene in the heterologous host. In opposition to this, sequence-based screening involves the techniques like polymerase chain reaction (PCR), southern hybridization to identify the desired gene with the help of DNA probes obtained from the known sample encoding for similar protein or enzyme of interest (National Research Council (US) Committee on a National Strategy for Biotechnology in Agriculture 1987). However, it overcomes the limitation of heterologous expression but demands for specialized databases for sequence analysis. Yet, both of these approaches are time-consuming and labour intensive considering the number of clones with desired traits. To surpass these limitations, Watanabe and his team members developed a new screening approach named SIGEX (Bull et al. 2000).

In the SIGEX, the expression is recorded by fluorescence-activated cell sorting (FACS) to detect the signal triggered by promoter region due to its interaction with the specific substrate. In general, it works on substrate-induced gene expression which get regulated by regulatory elements in close proximity, so it aids in screening the clone containing the substrate-inducible genes (Daugherty et al. 2000). For instance, catabolic genes expression of the inserted metagenomic DNA in clone was evaluated via co-expression of GFP (green fluorescent protein), in the presence of suitable substrate, and the clones showing the positive result were differentiated using FACS (Ekkers et al. 2012). Even though this method is economical and robust but requires the technical assistance to setup and analyze the data obtained through FACS. Furthermore, compound configuration screening is the latest approach which aid in identifying the targeted metabolites via chromatographic analysis supported by mass spectroscopy (Chong et al. 2018). In this screening relies on the ability of the clones to synthesize the new compounds and provide the diverse chromatographic peaks. Though this approach is effective but is laborious and cost-intensive. Each method has its pro and cons, but all these approach helps us unfolding the information about uncultured microbes that were previously unknown (Bisht and Panda 2013).

Due to the advancement in the various fields of science like bioinformatics, biotechnology, genomics and microbiology has led to development of metagenomic approach which is substantially helping in finding the novel enzymes.

7.4 Industrially-Important Enzymes Obtained Through Metagenomics

Metagenomics is predicted to be the most significant technological approach to bioprospect novel enzymes of industrial importance (Ahmad et al. 2019). Various industrially important enzymes like cellulases, amylases, lipases, proteases, xylanases etc. have been produced through metagenomics have been comprehended in Table 7.1. Moreover, major type of industrial important enzyme with their application has been depicted in Fig. 7.2. Some of the chief enzymes that have been unveiled from untapped resources have been discussed below:

Table 7.1 Comprehensive of list of the enzymes obtained by the metagenomic method in the years (2015–2020)

Approaches	Enzyme	Source	References
Sequence-based screening (SBS)	Acetyl Xylan esterase	Hot desert hypolith	Adesioye et al. (2018)
	Antibiotic-resistance enzymes	Atlantis II Deep Red Sea brine pool	Elbehery et al. (2017)
	β-Xylosidases	Hot spring soil	Sato et al. (2017)
	α-L arabinofuranosidase, β-glucosidase, β-xylosidase, and endo-1,4-β-xylanase	Porcupine microbiome (Erethizon dorsatum)	Thornbury et al. (2019)
	Bifunctional cellulose/ hemicellulose	Black goat rumen	Lee et al. (2018)
	Bilirubin-oxidizing enzyme	Wastewater treatment activated sludge	Kimura and Kamagata (2016)
	Carbohydrate-active enzyme	Antarctic tundra soil	Oh et al. (2019)
	Carbohydrate-active enzyme	Camel and goat intestinal tract	Al-Masaudi et al. (2019)
	Carbonic anhydrase	Active hydrothermal vent chimney	Fredslund et al. (2018)
	Carboxyl esterase	R. exoculata gill metagenome	Alcaide et al. (2015)
	Cellulases	Oil reservoir	Lewin et al. (2017)
	Epoxide hydrolases (EHs) Tomsk-LEH, CH55-LEH	Hot spring metagenomic library	Ferrandi et al. (2015)
	Esterase-active enzyme	Heated compost and hot spring	Leis et al. (2015)
	Glycoside hydrolase GH16 SCLam	Sugarcane soil metagenome	Alvarez et al. (2015)
	Hydrolases	Hot and other extreme environments	Wohlgemuth et al. (2018)
	Lignocellulose-degrading enzyme	Gut microbiome of the common black slug Arion ater	Joynson et al. (2017)
	Pectinase	Apple pomace-adapted compost	Zhou et al. (2017)
	Tauroursodeoxycholic acid biotransformation enzymes	Gut microbiome of black bears	Song et al. (2017)
	Thermostable amine transferases	Hot spring	Ferrandi et al. (2017)
	Toluene monooxygenase	Hydrocarbon-polluted sediment	Bouhajja et al. (2017)

(continued)

Table 7.1 (continued)

Approaches	Enzyme	Source	References
Functional-based screening	Alkaline protease	Mangrove sediment	Pessoa et al. (2017)
	β-Lactamase	Environment with high levels of biocide exposure	Salimraj et al. (2016)
	Carbamate degrading enzymes	Bovine rumen microbiome	Ufarté et al. (2017)
	Cellulases	Microorganisms associated with coral	Sousa et al. (2016)
	DNA endonuclease	Soil sample	Mtimka et al. (2020)
	Esterase	Sea sediment and hot spring microbial mat	Ranjan et al. (2018)
	Esterase	Deep-sea sediment	Ho et al. (2018)
	Esterase	Marine mud	Gao et al. (2016)
	Fe-Fe hydrogenase genes	Municipal wastewater treatment plant	Tomazetto et al. (2015)
	Fibrolytic enzymes	Indian crossbred cattle fed finger millet straw	Jose et al. (2017)
	Glycosyl hydrolase	Industrial soil bagasse collection site	Kanokratana et al. (2015)
	Glycosyl hydrolase	Mangrove soil	Mai et al. (2016)
	Hydrolases	Oil-impacted mangrove sediments	Ottoni et al. (2017)
	l-Asparaginase	Forest soil	Arjun et al. (2018)
	Laccase or laccase-like enzymes	Acidic bog soil	Ausec et al. (2017)
	Lignocellulases	Landfill sites	Ransom-Jones et al. (2017)
	Lignocellulase	Sugarcane bagasse and cow manure	Colombo et al. (2016)
	Lignin-transforming enzymes	Coal beds	Huo et al. (2018)
	Lipolytic biocatalysts	Global Ocean sampling dataset	Masuch et al. (2015)
	Peptidases	Yucatán underground water	Apolinar-Hernández et al. (2016)
	Rubisco	Hydrothermal vent fluid	Böhnke and Perner (2015)
	Esterase	Deep-sea sediment	Zhang et al. (2017)
	Serine and metalloproteases	Solid tannery waste	Verma and Sharma (2020)

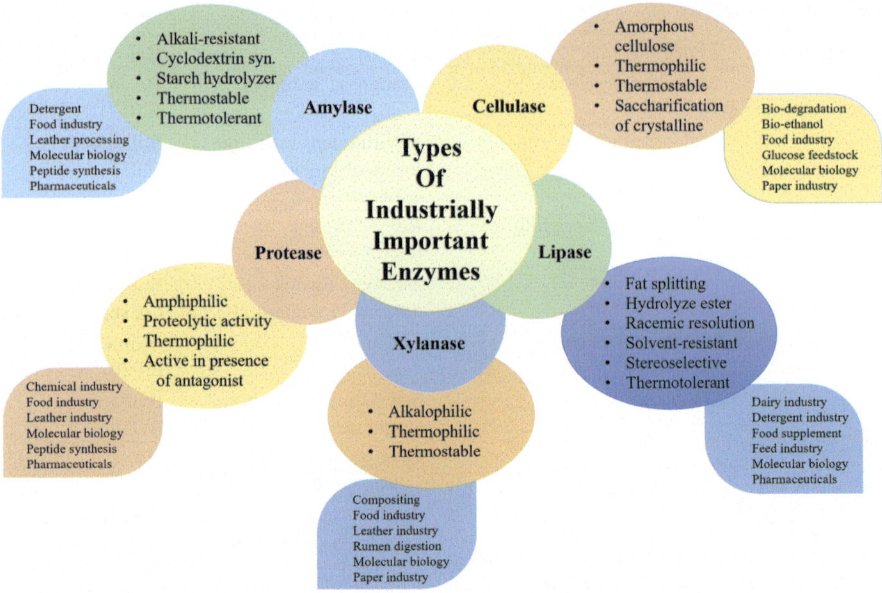

Fig. 7.2 Types of different industrial important enzymes and their application in various sectors

7.4.1 Amylases

Amylases belong to a group of enzymes called as glycoside hydrolases and are one of the widely used hydrolytic enzymes of great industrial significance. They have widespread application in various industries like fermentation, food, brewery, paper and textile industries. However, with the advancement in biotechnology, the application of amylases has spread to other sectors like analytical chemistry, medicinal, clinical etc. (de Souza and Magalhães 2010). They are involved in hydrolysis of starch to generate diversified products such as dextrin, small polymers comprising of glucose molecules etc. Researchers have identified amylases from metagenome of soil that possess the ability to retain 90% of their activity even under low temperature conditions and therefore, indicating the potential application of amylase in industries (Sharma et al. 2010).

7.4.2 Cellulases

Cellulases act on cellulosic polymers and break it down into its monomeric glucose units. They have widespread application in various industries like recycling of paper, processing of cotton, extraction of juices, as additives for animal feed, as a component in detergents etc. (Ostafe et al. 2014). Therefore, this ranks cellulase at third in worldwide enzyme production, by dollar volume. Metagenomics has been employed to unveil the potential novel cellulases from diverse environments such as rumen of animals, compost soils, soil of low temperature regions further building libraries of their metagenome and screening the active clones (Wang et al. 2016). Moreover, cellulases have captured significant attention owing to their ability to transform lignocellulosic biomass into biofuels like biodiesel, bioethanol etc. Hess et al. conducted a breakthrough study which led to the identification of 27,755 genes having significant similarity with either carbohydrate binding module or at least one of the catalytic domains present in enzyme (Mohanram et al. 2013). They generated extensive data to signify the potential of metagenomic sequencing of complex microbial communities in creating draft genomes of the non-culturable organisms and unveiling the genes encoding cellulase at massive scale that could potentially be significant in biomass depolymerization (Lemos et al. 2017). Using metagenomic approach, novel cellulases with diverse properties such as halotolerance, thermostability, acidophilic etc. have been identified and characterized from various sources like sugarcane soil, buffalo rumen etc. (Sahoo et al. 2018).

7.4.3 Proteases

Proteases are among the three largest groups of enzymes and most of them are derived from bacteria. Proteases serves various purposes in industries such as removal of stains, preparation of food, removal of biofilms, preparation of leather and so on (Singh and Bajaj 2017). Recently, many proteases have been identified using metagenomic-based techniques. Researchers have isolated protease that is alkali-tolerant and shows oxidant stability, therefore indicating its potential to be used for bleaching in detergent industries (Razzaq et al. 2019). Moreover, for application in detergent industry, two serine proteases, resistant towards detergent were isolated from desert by metagenomic approach by a group of researchers. Proteases also play an important role in diagnostic industries, thus to unveil the potential novel proteases from extreme environmental habitats, especially serine proteases and metalloproteases, there is a need to put extensive efforts in the field of metagenomics (Devi et al. 2016).

7.4.4 Lipases

Lipases form another sub-class of hydrolases involved in the hydrolysis of long-chain acyl glycerol such as trioleyl glycerol (standard substrate). Applications of lipases in industries are well-known such as in detergent industry, dairy industry, food industry, cosmetics, leather industry, chiral synthesis and wastewater treatment (Rogalskas et al. 1990). Since the discovery of new families of lipases from microbial communities, the search for novel lipases using metagenomic techniques remain unabated. Using these recent strategies, numerous lipase coding genes have been discovered and reported by scientists (Lopez-Lopez et al. 2014).

Peng et al. created a metagenomic library from marine sediments and identified a novel lipase that was stable under alkaline conditions and suggested its possible application in food industry in imparting a characteristic flavour and fragrance in production of milk fat flavour (Peng et al. 2014). In a study, Selvin et al. constructed a cosmid metagenomic library from marine sponge and using function-based metagenomic screening, they identified a novel halotolerant lipase. Moreover, it was found to be stable under wide range of pH and temperature conditions as well as could withstand the metals ions and harsh organic solvents therefore, making it suitable for various industrial applications (Selvin et al. 2012). Metagenomic approach, a culture independent strategy has the potential to untap genetic sources of lipases for diverse characteristics like enantioselectivity, extreme pH and temperature tolerance, substrate specificity etc. (Zhang et al. 2009).

7.4.5 Xylanases

Hemicellulose, a heterogenous polymer found in lignocellulosic biomass, is comprised of pentose sugars units and is the second most abundant polymers on earth after cellulose. Xylan (β-1,4-linked xylose) is the predominant sugar present in the complex matrix of hemicellulose. Endo-β-1,4-xylanase is the main enzyme involved in the degradation of xylan (Anwar et al. 2014). For efficient deconstruction of biomass and its bioconversion into biofuels, there is a necessity to develop efficient lignocellulolytic enzymes (Singh et al. 2019). The crosslinking of lignin, hemicellulose and cellulose by ether and ester linkages makes hemicellulose inaccessible for further processing. Therefore, it has triggered the search for novel xylanases that possess the ability to fractionate hemicellulose from lignin and cellulose (Bhardwaj et al. 2019). Moreover, xylanases ought to thermostable and tolerant towards extreme pH changes in order to serve well for biomass deconstruction. Verma et al. used metagenomic approach to construct metagenomic library from compost soil and identified a novel thermostable and alkali stable xylanase that could be potentially used for pulp bleaching in paper and pulp industry (Verma et al. 2013).

7.5 Limitations of Metagenomic Approach and Their Solutions

There are numerous limitations associated with metagenomics approach for identifying the new enzymes and certain measures are being taken to improve as well as reduce the time from enzyme identification to its application in industries (Alves et al. 2018). The major challenges are

(a) Presence of less number of genes encoding for enzyme of interest in metagenomic DNA,
(b) Limited number of enzyme that could work effectively in industrial conditions,
(c) Scarcity of the suitable substrate for functional screening,

(d) Low efficiency of screening procedure for rare activities,
(e) Low efficacy and performance of enzymes obtained by the approaches under artificial or induced conditions,
(f) High number of identified enzymes, which show no expression in expression vectors like *E. coli*
(g) Limited access to trustworthy bioinformatic tools for robust analysis of large amount of sequencing data and,
(h) Shortage of reliable prediction tools to predict the enzymatic activities on the basis of their coding sequence (Ferrer et al. 2016).

To overcome the above mentioned challenges, researchers are working on finding the solution which involves:

(a) Targeted screening procedure for isolating enzyme or gene encoding for enzyme of interest from massive clone libraries (Ufarté et al. 2015),
(b) Pre-enrichment under conditions similar to those required by processes or bio-transformations (Singh 2017),
(c) Identification and selection of genes as well as protein which express under specific conditions and on the specific substrate, that allows the selection of the enzyme with high activity (Chang et al. 2013),
(d) Selection of the enzyme with multiple application, high stability and wide specificity in varied conditions for various processes in industries (Alcaide et al. 2013),
(e) De novo synthesis of small molecules on demand which perform the similar function required by the industries for screening (Lim et al. 2013),
(f) Development of hosts and vectors for effective screening and expression of the gene of interest (Liebl et al. 2014),
(g) Incorporation of protein engineering with experimental and in silico studies for selection of new enzyme via metagenomic approach to develop more effective biotechnological variant (Alcaide et al. 2013),
(h) Construction of computational workflow to identify enzyme coding sequences on the basis of active-site modeling and structure-based function prediction with the help of bioinformatic platforms (Watson et al. 2007).

7.6 Conclusion

The onset of metagenomics has substantially improved the approach of exploring the new antibiotics, antibiotic resistance genes, novel microbial enzymes and transporters, etc. Even, the steps involved in the metagenomics approach like direct isolation of eDNA for various niches, then preparation of eDNA cloned libraries and later screening have exceptionally improved due to the availability of effective commercial kits. Moreover, various bioinformatics tools and software have been developed exclusively for metagenomic approach.

Lately, sequence-based investigation of metagenomes and DNA sequencing approach have surpassed the functional-based metagenomic investigations. The development of new bioinformatics tools has improved the annotation ability and now we can annotate thousands of metagenomes robustly. Whereas, function-based approach serves as the targeted approach but this approach cannot be neglected as it is crucial to identify novel enzymes and allow us to isolate as well as clone the gene of interest. And, then the cloned gene can be expressed in host as obtained from giga-bases of eDNA. Therefore, exploration of unexplored niches enables mining of the novel enzymes with desired characteristics, which aids in performing the biotechnological processes.

References

Adesioye FA, Makhalanyane TP, Vikram S, Sewell BT, Schubert WD, Cowana DA (2018) Structural characterization and directed evolution of a novel acetyl xylan esterase reveals thermostability determinants of the carbohydrate esterase 7 family. Appl Environ Microbiol 84:e02695-17. https://doi.org/10.1128/AEM.02695-17

Ahmad T, Singh RS, Gupta G, Sharma A, Kaur B (2019) Metagenomics in the search for industrial enzymes. In: Biomass, biofuels, biochemicals: advances in enzyme technology. Elsevier, Amsterdam, pp 419–451

Ahmed S, Zhou Z, Zhou J, Chen SQ (2016) Pharmacogenomics of drug metabolizing enzymes and transporters: relevance to precision medicine. Genomics Proteomics Bioinformatics 14:298–313

Alberts B, Johnson A, Lewis J, Raff M, Roberts K, Walter P (2002) Isolating, cloning, and sequencing DNA. In: Molecular biology of the cell, 4th edn. Garland Science, New York

Alcaide M, Tornés J, Stogios PJ, Xu X, Gertler C, Leo RDI, Bargiela R, Lafraya Á, Guazzaroni ME, López-Cortés N, Chernikova TN, Golyshina OV, Nechitaylo TY, Plumeier I, Pieper DH, Yakimov MM, Savchenko A, Golyshin PN, Ferrer M (2013) Single residues dictate the co-evolution of dual esterases: MCP hydrolases from the α/β hydrolase family. Biochem J 454:157–166. https://doi.org/10.1042/BJ20130552

Alcaide M, Tchigvintsev A, Martínez-Martínez M, Popovic A, Reva ON, Lafraya Á, Bargiela R, Nechitaylo TY, Matesanz R, Cambon-Bonavita MA, Jebbar M, Yakimov MM, Savchenko A, Golyshina OV, Yakunin AF, Golyshin PN, Ferrer M (2015) Identification and characterization of carboxyl esterases of gill chamber-associated microbiota in the deep-sea shrimp Rimicaris exoculata by using functional metagenomics. Appl Environ Microbiol 81:2125–2136. https://doi.org/10.1128/AEM.03387-14

Al-Masaudi S, El Kaoutari A, Drula E, Redwan EM, Lombard V, Henrissat B (2019) A metagenomics investigation of carbohydrate-active enzymes along the goat and camel intestinal tract. Int Microbiol 22:429–435. https://doi.org/10.1007/s10123-019-00068-2

Alvarez TM, Liberato MV, Cairo JPLF, Paixão DAA, Campos BM, Ferreira MR, Almeida RF, Pereira IO, Bernardes A, Ematsu GCG, Chinaglia M, Polikarpov I, de Oliveira NM, Squina FM (2015) A novel member of GH16 family derived from sugarcane soil metagenome. Appl Biochem Biotechnol 177:304–317. https://doi.org/10.1007/s12010-015-1743-7

Alves LF, Westmann CA, Lovate GL, de Siqueira GMV, Borelli TC, Guazzaroni M-E (2018) Metagenomic approaches for understanding new concepts in microbial science. Int J Genomics 2018:2312987. https://doi.org/10.1155/2018/2312987

Anand P, Chopra RS, Dhanjal DS, Chopra C (2019) Isolation and characterization of microbial diversity of soil of Dhanbad coal mines using molecular approach. Res J Pharm Technol 12:1137–1140. https://doi.org/10.5958/0974-360X.2019.00187.2

Anwar Z, Gulfraz M, Irshad M (2014) Agro-industrial lignocellulosic biomass a key to unlock the future bio-energy: a brief review. J Radiat Res Appl Sci 7:163–173. https://doi.org/10.1016/j.jrras.2014.02.003

Apolinar-Hernández MM, Peña-Ramírez YJ, Pérez-Rueda E, Canto-Canché BB, De los Santos-Briones C, O'Connor-Sánchez A (2016) Identification and in silico characterization of two novel genes encoding peptidases S8 found by functional screening in a metagenomic library of Yucatán underground water. Gene 593:154–161. https://doi.org/10.1016/j.gene.2016.08.009

Arjun JK, Aneesh BP, Kavitha T, Harikrishnan K (2018) Characterization of a novel asparaginase from soil metagenomic libraries generated from forest soil. Biotechnol Lett 40:343–348. https://doi.org/10.1007/s10529-017-2470-7

Ausec L, Berini F, Casciello C, Cretoiu MS, van Elsas JD, Marinelli F, Mandic-Mulec I (2017) The first acidobacterial laccase-like multicopper oxidase revealed by metagenomics shows high salt and thermo-tolerance. Appl Microbiol Biotechnol 101:6261–6276. https://doi.org/10.1007/s00253-017-8345-y

Bajpai B (2014) High capacity vectors. In: Advances in biotechnology. Springer, New Delhi, pp 1–10

Baweja M, Nain L, Kawarabayasi Y, Shukla P (2016) Current technological improvements in enzymes toward their biotechnological applications. Front Microbiol 7:965

Bhardwaj N, Kumar B, Verma P (2019) A detailed overview of xylanases: an emerging biomolecule for current and future prospective. Bioresour Bioprocess 6:1–36

Bisht SS, Panda AK (2013) DNA sequencing: methods and applications. In: Advances in biotechnology. Springer, New Delhi, pp 11–23

Böhnke S, Perner M (2015) A function-based screen for seeking RubisCO active clones from metagenomes: novel enzymes influencing RubisCO activity. ISME J 9:735–745. https://doi.org/10.1038/ismej.2014.163

Bouhajja E, McGuire M, Liles MR, Bataille G, Agathos SN, George IF (2017) Identification of novel toluene monooxygenase genes in a hydrocarbon-polluted sediment using sequence- and function-based screening of metagenomic libraries. Appl Microbiol Biotechnol 101:797–808. https://doi.org/10.1007/s00253-016-7934-5

Bull AT, Ward AC, Goodfellow M (2000) Search and discovery strategies for biotechnology: the paradigm shift. Microbiol Mol Biol Rev 64:573–606. https://doi.org/10.1128/mmbr.64.3.573-606.2000

Chang C, Sustarich J, Bharadwaj R, Chandrasekaran A, Adams PD, Singh AK (2013) Droplet-based microfluidic platform for heterogeneous enzymatic assays. Lab Chip 13:1817–1822. https://doi.org/10.1039/c3lc41418c

Cheng J, Romantsov T, Engel K, Doxey AC, Rose DR, Neufeld JD, Charles TC (2017) Functional metagenomics reveals novel β-galactosidases not predictable from gene sequences. PLoS One 12:e0172545–e0172545. https://doi.org/10.1371/journal.pone.0172545

Chong YK, Ho CC, Leung SY, Lau SKP, Woo PCY (2018) Clinical mass spectrometry in the bioinformatics era: a Hitchhiker's guide. Comput Struct Biotechnol J 16:316–334

Colombo LT, de Oliveira MNV, Carneiro DG, de Souza RA, Alvim MCT, dos Santos JC, da Silva CC, Vidigal PMP, da Silveira WB, Passos FML (2016) Applying functional metagenomics to search for novel lignocellulosic enzymes in a microbial consortium derived from a thermophilic composting phase of sugarcane bagasse and cow manure. Antonie Van Leeuwenhoek 109:1217–1233. https://doi.org/10.1007/s10482-016-0723-4

Coughlan LM, Cotter PD, Hill C, Alvarez-Ordóñez A (2015) Biotechnological applications of functional metagenomics in the food and pharmaceutical industries. Front Microbiol 6:672

Daugherty PS, Iverson BL, Georgiou G (2000) Flow cytometric screening of cell-based libraries. J Immunol Methods 243:211–227

de Souza PM, Magalhães PDO (2010) Application of microbial α-amylase in industry - a review. Braz J Microbiol 41:850–861

DeCastro ME, Rodríguez-Belmonte E, González-Siso MI (2016) Metagenomics of thermophiles with a focus on discovery of novel thermozymes. Front Microbiol 7:1521

Devi SG, Fathima AA, Sanitha M, Iyappan S, Curtis WR, Ramya M (2016) Expression and characterization of alkaline protease from the metagenomic library of tannery activated sludge. J Biosci Bioeng 122:694–700. https://doi.org/10.1016/j.jbiosc.2016.05.012

Dhanjal DS, Sharma D (2018) Microbial metagenomics for industrial and environmental bioprospecting: the unknown envoy. In: Microbial bioprospecting for sustainable development. Springer, Singapore, pp 327–352

Dhanjal DS, Chopra C, Anand P, Chopra RS (2017) Accessing the microbial diversity of sugarcane fields from Gujjarwal village, Ludhiana and their molecular identification. Res J Pharm Technol 10:3439–3442. https://doi.org/10.5958/0974-360X.2017.00612.6

Ekkers DM, Cretoiu MS, Kielak AM, Van Elsas JD (2012) The great screen anomaly-a new frontier in product discovery through functional metagenomics. Appl Microbiol Biotechnol 93:1005–1020

Elbehery AHA, Leak DJ, Siam R (2017) Novel thermostable antibiotic resistance enzymes from the Atlantis II Deep Red Sea brine pool. Microb Biotechnol 10:189–202. https://doi.org/10.1111/1751-7915.12468

Engqvist MKM, Rabe KS (2019) Applications of protein engineering and directed evolution in plant research. Plant Physiol 179:907–917. https://doi.org/10.1104/pp.18.01534

Escobar-Zepeda A, De León AVP, Sanchez-Flores A (2015) The road to metagenomics: from microbiology to DNA sequencing technologies and bioinformatics. Front Genet 6:348

Ferrandi EE, Sayer C, Isupov MN, Annovazzi C, Marchesi C, Iacobone G, Peng X, Bonch-Osmolovskaya E, Wohlgemuth R, Littlechild JA, Monti D (2015) Discovery and characterization of thermophilic limonene-1, 2-epoxide hydrolases from hot spring metagenomic libraries. FEBS J 282:2879–2894. https://doi.org/10.1111/febs.13328

Ferrandi EE, Previdi A, Bassanini I, Riva S, Peng X, Monti D (2017) Novel thermostable amine transferases from hot spring metagenomes. Appl Microbiol Biotechnol 101:4963–4979. https://doi.org/10.1007/s00253-017-8228-2

Ferrer M, Beloqui A, Timmis KN, Golyshin PN (2008) Metagenomics for mining new genetic resources of microbial communities. J Mol Microbiol Biotechnol 16:109–123

Ferrer M, Martínez-Martínez M, Bargiela R, Streit WR, Golyshina OV, Golyshin PN (2016) Estimating the success of enzyme bioprospecting through metagenomics: current status and future trends. Microb Biotechnol 9:22–34

Fredslund F, Borchert MS, Poulsen JCN, Mortensen SB, Perner M, Streit WR, Lo Leggio L (2018) Structure of a hyperthermostable carbonic anhydrase identified from an active hydrothermal vent chimney. Enzym Microb Technol 114:48–54. https://doi.org/10.1016/j.enzmictec.2018.03.009

Gabor EM, De Vries EJ, Janssen DB (2003) Efficient recovery of environmental DNA for expression cloning by indirect extraction methods. FEMS Microbiol Ecol 44:153–163. https://doi.org/10.1016/S0168-6496(02)00462-2

Gao W, Wu K, Chen L, Fan H, Zhao Z, Gao B, Wang H, Wei D (2016) A novel esterase from a marine mud metagenomic library for biocatalytic synthesis of short-chain flavor esters. Microb Cell Factories 15:1–12. https://doi.org/10.1186/s12934-016-0435-5

Garza DR, Dutilh BE (2015) From cultured to uncultured genome sequences: Metagenomics and modeling microbial ecosystems. Cell Mol Life Sci 72:4287–4308

Gu W, Miller S, Chiu CY (2019) Clinical metagenomic next-generation sequencing for pathogen detection. Annu Rev Pathol Mech Dis 14:319–338. https://doi.org/10.1146/annurev-pathmechdis-012418-012751

Gurung N, Ray S, Bose S, Rai V (2013) A broader view: microbial enzymes and their relevance in industries, medicine, and beyond. Biomed Res Int 2013:329121. https://doi.org/10.1155/2013/329121

Handelsman J (2004) Metagenomics: application of genomics to uncultured microorganisms. Microbiol Mol Biol Rev 68:669–685. https://doi.org/10.1128/mmbr.68.4.669-685.2004

Ho JCH, Pawar SV, Hallam SJ, Yadav VG (2018) An improved whole-cell biosensor for the discovery of lignin-transforming enzymes in functional metagenomic screens. ACS Synth Biol 7:392–398. https://doi.org/10.1021/acssynbio.7b00412

Huo YY, Jian SL, Cheng H, Rong Z, Cui HL, Xu XW (2018) Two novel deep-sea sediment metagenome-derived esterases: residue 199 is the determinant of substrate specificity and preference. Microb Cell Factories 17:1–12. https://doi.org/10.1186/s12934-018-0864-4

International Energy Agency (2020) India 2020 - energy policy review, India

Iqbal HA, Craig JW, Brady SF (2014) Antibacterial enzymes from the functional screening of metagenomic libraries hosted in Ralstonia metallidurans. FEMS Microbiol Lett 354:19–26

Jose VL, Appoothy T, More RP, Arun AS (2017) Metagenomic insights into the rumen microbial fibrolytic enzymes in Indian crossbred cattle fed finger millet straw. AMB Express 7:1–11. https://doi.org/10.1186/s13568-016-0310-0

Joynson R, Pritchard L, Osemwekha E, Ferry N (2017) Metagenomic analysis of the gut microbiome of the common black slug arion ater in search of novel lignocellulose degrading enzymes. Front Microbiol 8:2181. https://doi.org/10.3389/fmicb.2017.02181

Kanokratana P, Eurwilaichitr L, Pootanakit K, Champreda V (2015) Identification of glycosyl hydrolases from a metagenomic library of microflora in sugarcane bagasse collection site and their cooperative action on cellulose degradation. J Biosci Bioeng 119:384–391. https://doi.org/10.1016/j.jbiosc.2014.09.010

Kimura N (2014) Metagenomic approaches to understanding phylogenetic diversity in quorum sensing. Virulence 5:433–442

Kimura N, Kamagata Y (2016) A thermostable bilirubin-oxidizing enzyme from activated sludge isolated by a metagenomic approach. Microbes Environ 31:435–441. https://doi.org/10.1264/jsme2.ME16106

Kotik M (2009) Novel genes retrieved from environmental DNA by polymerase chain reaction: current genome-walking techniques for future metagenome applications. J Biotechnol 144:75–82

Kunin V, Copeland A, Lapidus A, Mavromatis K, Hugenholtz P (2008) A bioinformatician's guide to metagenomics. Microbiol Mol Biol Rev 72:557–578. https://doi.org/10.1128/mmbr.00009-08

Lam KN, Cheng J, Engel K, Neufeld JD, Charles TC (2015) Current and future resources for functional metagenomics. Front Microbiol 6:1196. https://doi.org/10.3389/fmicb.2015.01196

Lee MH, Lee S-W (2013) Bioprospecting potential of the soil metagenome: novel enzymes and bioactivities. Genomics Inform 11:114. https://doi.org/10.5808/gi.2013.11.3.114

Lee KT, Toushik SH, Baek JY, Kim JE, Lee JS, Kim KS (2018) Metagenomic mining and functional characterization of a novel KG51 bifunctional cellulase/hemicellulase from black goat rumen. J Agric Food Chem 66:9034–9041. https://doi.org/10.1021/acs.jafc.8b01449

Leis B, Angelov A, Mientus M, Li H, Pham VTT, Lauinger B, Bongen P, Pietruszka J, GonÃ§alves LG, Santos H, Liebl W (2015) Identification of novel esterase-active enzymes from hot environments by use of the host bacterium Thermus thermophilus. Front Microbiol 6:275. https://doi.org/10.3389/fmicb.2015.00275

Lemos LN, Pereira RV, Quaggio RB, Martins LF, Moura LMS, da Silva AR, Antunes LP, da Silva AM, Setubal JC (2017) Genome-centric analysis of a thermophilic and cellulolytic bacterial consortium derived from composting. Front Microbiol 8:644. https://doi.org/10.3389/fmicb.2017.00644

Lewin A, Zhou J, Pham VTT, Haugen T, El Zeiny M, Aarstad O, Liebl W, Wentzel A, Liles MR (2017) Novel archaeal thermostable cellulases from an oil reservoir metagenome. AMB Express 7:183. https://doi.org/10.1186/s13568-017-0485-z

Liebl W, Angelov A, Juergensen J, Chow J, Loeschcke A, Drepper T, Classen T, Pietruzska J, Ehrenreich A, Streit WR, Jaeger KE (2014) Alternative hosts for functional (meta)genome analysis. Appl Microbiol Biotechnol 98:8099–8109

Lim J, Vrignon J, Gruner P, Karamitros CS, Konrad M, Baret JC (2013) Ultra-high throughput detection of single cell β-galactosidase activity in droplets using micro-optical lens array. Appl Phys Lett 103:203704. https://doi.org/10.1063/1.4830046

Lopez-Lopez O, Cerdan M, Siso M (2014) New extremophilic lipases and esterases from metagenomics. Curr Protein Pept Sci 15:445–455. https://doi.org/10.2174/1389203715666140228153801

Louca S, Mazel F, Doebeli M, Parfrey LW (2019) A census-based estimate of earth's bacterial and archaeal diversity. PLoS Biol 17:e3000106. https://doi.org/10.1371/journal.pbio.3000106

Mai Z, Su H, Zhang S (2016) Isolation and characterization of a glycosyl hydrolase family 16 β-agarase from a mangrove soil metagenomic library. Int J Mol Sci 17:1360. https://doi.org/10.3390/ijms17081360

Malla MA, Dubey A, Kumar A, Yadav S, Hashem A, Allah EFA (2019) Exploring the human microbiome: the potential future role of next-generation sequencing in disease diagnosis and treatment. Front Immunol 10:2868

Maria-Solano MA, Serrano-Hervás E, Romero-Rivera A, Iglesias-Fernández J, Osuna S (2018) Role of conformational dynamics in the evolution of novel enzyme function. Chem Commun 54:6622–6634. https://doi.org/10.1039/c8cc02426j

Markel U, Essani KD, Besirlioglu V, Schiffels J, Streit WR, Schwaneberg U (2020) Advances in ultrahigh-throughput screening for directed enzyme evolution. Chem Soc Rev 49:233–262

Masuch T, Kusnezowa A, Nilewski S, Bautista JT, Kourist R, Leichert LI (2015) A combined bioinformatics and functional metagenomics approach to discovering lipolytic biocatalysts. Front Microbiol 6:1110. https://doi.org/10.3389/fmicb.2015.01110

McNichol J, Stryhanyuk H, Sylva SP, Thomas F, Musat N, Seewald JS, Sievert SM (2018) Primary productivity below the seafloor at deep-sea hot springs. Proc Natl Acad Sci U S A 115:6756–6761. https://doi.org/10.1073/pnas.1804351115

Meier MJ, Paterson ES, Lambert IB (2016) Use of substrate-induced gene expression in metagenomic analysis of an aromatic hydrocarbon-contaminated soil. Appl Environ Microbiol 82:897–909. https://doi.org/10.1128/AEM.03306-15

Mohanram S, Amat D, Choudhary J, Arora A, Nain L (2013) Novel perspectives for evolving enzyme cocktails for lignocellulose hydrolysis in biorefineries. Sustain Chem Process 1:15. https://doi.org/10.1186/2043-7129-1-15

Morimoto S, Fujii T (2009) A new approach to retrieve full lengths of functional genes from soil by PCR-DGGE and metagenome walking. Appl Microbiol Biotechnol 83:389–396. https://doi.org/10.1007/s00253-009-1992-x

Mtimka S, Pillay P, Rashamuse K, Gildenhuys S, Tsekoa TL (2020) Functional screening of a soil metagenome for DNA endonucleases by acquired resistance to bacteriophage infection. Mol Biol Rep 47:353–361. https://doi.org/10.1007/s11033-019-05137-3

Mukherjee A, Chettri B, Langpoklakpam JS, Basak P, Prasad A, Mukherjee AK, Bhattacharyya M, Singh AK, Chattopadhyay D (2017) Bioinformatic approaches including predictive metagenomic profiling reveal characteristics of bacterial response to petroleum hydrocarbon contamination in diverse environments. Sci Rep 7:1108. https://doi.org/10.1038/s41598-017-01126-3

Mukherjee D, Singh S, Kumar M, Kumar V, Datta S, Dhanjal DS (2018) Fungal biotechnology: role and aspects. In: Fungi and their role in sustainable development: current perspective. Springer, Singapore, pp 91–103

National Research Council (US) Committee on a National Strategy for Biotechnology in Agriculture (1987) Gene transfer methods applicable to agricultural organisms

National Research Council (US) Committee on Noneconomic and Economic Value of Biodiversity (1999) The values of biodiversity

Ngara TR, Zhang H (2018) Recent advances in function-based metagenomic screening. Genomics Proteomics Bioinformatics 16:405–415

Oh HN, Park D, Seong HJ, Kim D, Sul WJ (2019) Antarctic tundra soil metagenome as useful natural resources of cold-active lignocelluolytic enzymes. J Microbiol 57:865–873. https://doi.org/10.1007/s12275-019-9217-1

Ostafe R, Prodanovic R, Lloyd Ung W, Weitz DA, Fischer R (2014) A high-throughput cellulase screening system based on droplet microfluidics. Biomicrofluidics 8:041102. https://doi.org/10.1063/1.4886771

Ottoni JR, Cabral L, de Sousa STP, Júnior GVL, Domingos DF, Soares Junior FL, da Silva MCP, Marcon J, Dias ACF, de Melo IS, de Souza AP, Andreote FD, de Oliveira VM (2017) Functional metagenomics of oil-impacted mangrove sediments reveals high abundance of hydrolases of biotechnological interest. World J Microbiol Biotechnol 33:1–13. https://doi.org/10.1007/s11274-017-2307-5

Oulas A, Pavloudi C, Polymenakou P, Pavlopoulos GA, Papanikolaou N, Kotoulas G, Arvanitidis C, Iliopoulos I (2015) Metagenomics: tools and insights for analyzing next-generation sequencing data derived from biodiversity studies. Bioinform Biol Insights 9:75–88. https://doi.org/10.4137/BBI.S12462

Peng Q, Wang X, Shang M, Huang J, Guan G, Li Y, Shi B (2014) Isolation of a novel alkaline-stable lipase from a metagenomic library and its specific application for milkfat flavor production. Microb Cell Factories 13:1. https://doi.org/10.1186/1475-2859-13-1

Pessoa TB, Rezende RP, Marques ED, Pirovani CP, dos Santos TF, dos Santos Gonçalves AC, Romano CC, Dotivo NC, Freitas AC, Salay LC, Dias JC (2017) Metagenomic alkaline protease from mangrove sediment. J Basic Microbiol 57:962–973. https://doi.org/10.1002/jobm.201700159

Popovic A, Hai T, Tchigvintsev A, Hajighasemi M, Nocek B, Khusnutdinova AN, Brown G, Glinos J, Flick R, Skarina T, Chernikova TN, Yim V, Brüls T, Le Paslier D, Yakimov MM, Joachimiak A, Ferrer M, Golyshina OV, Savchenko A, Golyshin PN, Yakunin AF (2017) Activity screening of environmental metagenomic libraries reveals novel carboxylesterase families. Sci Rep 7:1–15. https://doi.org/10.1038/srep44103

Ranjan R, Yadav MK, Suneja G, Sharma R (2018) Discovery of a diverse set of esterases from hot spring microbial mat and sea sediment metagenomes. Int J Biol Macromol 119:572–581. https://doi.org/10.1016/j.ijbiomac.2018.07.170

Ransom-Jones E, McCarthy AJ, Haldenby S, Doonan J, McDonald JE (2017) Lignocellulose-degrading microbial communities in landfill sites represent a repository of unexplored biomass-degrading diversity. mSphere 2:e00300-17. https://doi.org/10.1128/msphere.00300-17

Raveendran S, Parameswaran B, Ummalyma SB, Abraham A, Mathew AK, Madhavan A, Rebello S, Pandey A (2018) Applications of microbial enzymes in food industry. Food Technol Biotechnol 56:16–30

Razzaq A, Shamsi S, Ali A, Ali Q, Sajjad M, Malik A, Ashraf M (2019) Microbial proteases applications. Front Bioeng Biotechnol 7:110

Robinson PK (2015) Enzymes: principles and biotechnological applications. Essays Biochem 59:1–41. https://doi.org/10.1042/BSE0590001

Rogalskas E, Ransac S, Vergers R (1990) Stereoselectivity of lipases II. Stereoselective hydrolysis of triglycerides by gastric and pancreatic lipases. J Biol Chem 265(33):20271–20276

Roopnarain A, Mukhuba M, Adeleke R, Moeletsi M (2017) Biases during DNA extraction affect bacterial and archaeal community profile of anaerobic digestion samples. 3. Biotech 7:375. https://doi.org/10.1007/s13205-017-1009-x

Rosen G, Sokhansanj B, Polikar R, Bruns M, Russell J, Garbarine E, Essinger S, Yok N (2009) Signal processing for metagenomics: extracting information from the soup. Curr Genomics 10:493–510. https://doi.org/10.2174/138920209789208255

Sahoo K, Sahoo RK, Gaur M, Subudhi E (2018) Isolation of cellulase genes from thermophilies: a novel approach toward new gene discovery. In: New and future developments in microbial biotechnology and bioengineering: microbial genes biochemistry and applications. Elsevier, Amsterdam, pp 151–169

Salimraj R, Zhang L, Hinchliffe P, Wellington EMH, Brem J, Schofield CJ, Gaze WH, Spencer J (2016) Structural and biochemical characterization of Rm3, a subclass B3 metallo-β-lactamase identified from a functional metagenomic study. Antimicrob Agents Chemother 60:5828–5840. https://doi.org/10.1128/AAC.00750-16

Sarangi M, Chopra C, Usman YA, Dhanjal DS, Chopra RS (2019) Accessing genetic diversity and phylogenetic analysis of microbial population of soil from Hygam Wetland of Kashmir Valley. Res J Pharm Technol 12:2323. https://doi.org/10.5958/0974-360x.2019.00391.3

Sato M, Suda M, Okuma J, Kato T, Hirose Y, Nishimura A, Kondo Y, Shibata D (2017) Isolation of highly thermostable b-xylosidases from a hot spring soil microbial community using a metagenomic approach. DNA Res 24:649–656. https://doi.org/10.1093/dnares/dsx032

Selvin J, Kennedy J, Lejon DPH, Kiran GS, Dobson ADW (2012) Isolation identification and biochemical characterization of a novel halo-tolerant lipase from the metagenome of the marine sponge Haliclona simulans. Microb Cell Factories 11:72. https://doi.org/10.1186/1475-2859-11-72

Shamim K, Sharma J, Dubey SK (2017) Rapid and efficient method to extract metagenomic DNA from estuarine sediments. 3 Biotech 7:182. https://doi.org/10.1007/s13205-017-0846-y

Sharma S, Khan FG, Qazi GN (2010) Molecular cloning and characterization of amylase from soil metagenomic library derived from northwestern Himalayas. Appl Microbiol Biotechnol 86:1821–1828. https://doi.org/10.1007/s00253-009-2404-y

Simon C, Daniel R (2011) Metagenomic analyses: past and future trends. Appl Environ Microbiol 77:1153–1161

Singh R (2017) Microbial biotransformation: a process for chemical alterations. J Bacteriol Mycol Open Access 4:00085. https://doi.org/10.15406/jbmoa.2017.04.00085

Singh S, Bajaj BK (2017) Potential application spectrum of microbial proteases for clean and green industrial production. Energy Ecol Environ 2:370–386

Singh R, Kumar M, Mittal A, Mehta PK (2016) Microbial enzymes: industrial progress in 21st century. 3 Biotech 6:174

Singh S, Sidhu GK, Kumar V, Dhanjal DS, Datta S, Singh J (2019) Fungal xylanases: sources, types, and biotechnological applications. Springer, Cham, pp 405–428

Song C, Wang B, Tan J, Zhu L, Lou D (2017) Discovery of tauroursodeoxycholic acid biotransformation enzymes from the gut microbiome of black bears using metagenomics. Sci Rep 7:1–8. https://doi.org/10.1038/srep45495

Sousa FMO, Moura SR, Quinto CA, Dias JCT, Pirovani CP, Rezende RP (2016) Functional screening for cellulolytic activity in a metagenomic fosmid library of microorganisms associated with coral. Genet Mol Res 15:gmr-15048770. https://doi.org/10.4238/gmr.15048770

Thornbury M, Sicheri J, Slaine P, Getz IDLJ, Finlayson-Trick E, Cook J, Guinard IDC, Boudreau N, JakemanID D, Rohde J, McCormickID C (2019) Characterization of novel lignocellulose-degrading enzymes from the porcupine microbiome using synthetic metagenomics. PLoS One 14:e0209221. https://doi.org/10.1371/journal.pone.0209221

Tomazetto G, Wibberg D, Schlüter A, Oliveira VM (2015) New FeFe-hydrogenase genes identified in a metagenomic fosmid library from a municipal wastewater treatment plant as revealed by high-throughput sequencing. Res Microbiol 166:9–19. https://doi.org/10.1016/j.resmic.2014.11.002

Uchiyama T, Miyazaki K (2010) Product-induced gene expression, a product-responsive reporter assay used to screen metagenomic libraries for enzyme-encoding genes. Appl Environ Microbiol 76:7029–7035. https://doi.org/10.1128/AEM.00464-10

Uchiyama T, Watanabe K (2008) Substrate-induced gene expression (SIGEX) screening of metagenome libraries. Nat Protoc 3:1202–1212. https://doi.org/10.1038/nprot.2008.96

Ufarté L, Potocki-Veronese G, Laville É (2015) Discovery of new protein families and functions: new challenges in functional metagenomics for biotechnologies and microbial ecology. Front Microbiol 6:563

Ufarté L, Laville E, Duquesne S, Morgavi D, Robe P, Klopp C, Rizzo A, Pizzut-Serin S, Potocki-Veronese G (2017) Discovery of carbamate degrading enzymes by functional metagenomics. PLoS One 12:e0189201. https://doi.org/10.1371/journal.pone.0189201

Uhlik O, Leewis MC, Strejcek M, Musilova L, Mackova M, Leigh MB, Macek T (2013) Stable isotope probing in the metagenomics era: a bridge towards improved bioremediation. Biotechnol Adv 31:154–165

Valetti F, Gilardi G (2013) Improvement of biocatalysts for industrial and environmental purposes by saturation mutagenesis. Biomol Ther 3:778–811

Van Der Helm E, Genee HJ, Sommer MOA (2018) The evolving interface between synthetic biology and functional metagenomics. Nat Chem Biol 14:752–759

Verma SK, Sharma PC (2020) NGS-based characterization of microbial diversity and functional profiling of solid tannery waste metagenomes. Genomics 112(4):2903–2913. https://doi.org/10.1016/j.ygeno.2020.04.002

Verma D, Kawarabayasi Y, Miyazaki K, Satyanarayana T (2013) Cloning, expression and characteristics of a novel alkalistable and thermostable xylanase encoding gene (Mxyl) retrieved from compost-soil metagenome. PLoS One 8:e52459. https://doi.org/10.1371/journal.pone.0052459

Vester JK, Glaring MA, Stougaard P (2015) Improved cultivation and metagenomics as new tools for bioprospecting in cold environments. Extremophiles 19:17–29

Wang C, Dong D, Wang H, Müller K, Qin Y, Wang H, Wu W (2016) Metagenomic analysis of microbial consortia enriched from compost: new insights into the role of actinobacteria in lignocellulose decomposition. Biotechnol Biofuels 9:22. https://doi.org/10.1186/s13068-016-0440-2

Watson JD, Sanderson S, Ezersky A, Savchenko A, Edwards A, Orengo C, Joachimiak A, Laskowski RA, Thornton JM (2007) Towards fully automated structure-based function prediction in structural genomics: a case study. J Mol Biol 367:1511–1522. https://doi.org/10.1016/j.jmb.2007.01.063

Williamson LL, Borlee BR, Schloss PD, Guan C, Allen HK, Handelsman J (2005) Intracellular screen to identify metagenomic clones that induce or inhibit a quorum-sensing biosensor. Appl Environ Microbiol 71:6335–6344. https://doi.org/10.1128/AEM.71.10.6335-6344.2005

Wohlgemuth R, Littlechild J, Monti D, Schnorr K, van Rossum T, Siebers B, Menzel P, Kublanov IV, Rike AG, Skretas G, Szabo Z, Peng X, Young MJ (2018) Discovering novel hydrolases from hot environments. Biotechnol Adv 36:2077–2100

Wójcik M, Telzerow A, Quax WJ, Boersma YL (2015) High-throughput screening in protein engineering: recent advances and future perspectives. Int J Mol Sci 16:24918–24945

Woodley JM (2013) Protein engineering of enzymes for process applications. Curr Opin Chem Biol 17:310–316

Yadav D, Tanveer A, Yadav S (2019) Metagenomics for novel enzymes: a current perspective. In: Environmental contaminants: ecological implications and management. Springer, Singapore, pp 137–162

Zhang Y, Pengjun S, Wanli L, Kun M, Yingguo B, Guozeng W, Zhichun Z, Bin Y (2009) Lipase diversity in glacier soil based on analysis of metagenomic DNA fragments and cell culture. J Microbiol Biotechnol 19:888–897. https://doi.org/10.4014/jmb.0812.695

Zhang Y, Hao J, Zhang Y-Q, Chen X-L, Xie B-B, Shi M, Zhou B-C, Zhang Y-Z, Li P-Y (2017) Identification and characterization of a novel salt-tolerant esterase from the deep-sea sediment of the South China Sea. Front Microbiol 08:441. https://doi.org/10.3389/fmicb.2017.00441

Zhou J, He Z, Yang Y, Deng Y, Tringe SG, Alvarez-Cohen L (2015) High-throughput metagenomic technologies for complex microbial community analysis: open and closed formats. MBio 6:e02288-14. https://doi.org/10.1128/mBio.02288-14

Zhou M, Guo P, Wang T, Gao L, Yin H, Cai C, Gu J, Lü X (2017) Metagenomic mining pectinolytic microbes and enzymes from an apple pomace-adapted compost microbial community. Biotechnol Biofuels 10:198. https://doi.org/10.1186/s13068-017-0885-y

Metagenomics and Drug-Discovery 8

Bhupender Singh and Ayan Roy

Abstract

The emergence of High-throughput sequencing and its implication in the analysis of microbial population has introduced a new area of scientific research-metagenomics. Metagenomic analysis has brought a revolution in various fields of biological research, notably drug discovery. The term refers to the collective examination of genome analysis of unculturable microbial communities residing in a specific type of environmental condition or niche. A comprehensive investigation of the microbial diversity of unexploited areas with the aid of molecular biology and High-throughput sequencing technologies has opened the floodgates to explore and profile varieties of novel bioactive metabolites and potential antibiotics that promise to be of immense gravity for the pharmaceutical sector. High-throughput sequencing has accelerated the process of metabolite identification from metagenomic samples. Several bioactive metabolites have been obtained from metagenomic samples with immense therapeutic potential. Some examples include malacidin, fluoroquinolone, minimide and erdacin. In this chapter, major benchmark studies executed on the pharmacologically significant bioactive metabolites, extracted from metagenomic samples, have been discussed elaborately. An extensive review has also been conducted on several specialised bioinformatics-based pipelines frequently employed for the purpose. The present approach also aims at highlighting the major unexplored areas of drug discovery from metagenomic samples and associated metabolites—a hidden treasure for the pharmaceutical sector.

B. Singh · A. Roy (✉)
School of Bioengineering and Biosciences, Lovely Professional University, Phagwara, Punjab, India
e-mail: ayanroy.24373@lpu.co.in

R. S. Chopra et al. (eds.), *Metagenomics: Techniques, Applications, Challenges and Opportunities*, https://doi.org/10.1007/978-981-15-6529-8_8

Keywords

Metagenomics · Drug discovery · High-throughput sequencing · Metabolites · Antibiotics

8.1 Introduction

The recent advancement in microbiology has directed researchers from the respective discipline to explore microbes differently and insightfully (Arnold et al. 2016). The finding that most microbes cannot be cultured acted as a catalyst to alter the before used dynamics to study microbial populations (Stewart 2012). The knowledge of the impact of microbes on the humans and the environment has led the microbiologists to develop strategies for examining the uncultured microorganisms (Handelsman 2004). The urge to study the evolutionary and functional characteristics of the uncultured microbial diversity has introduced the multidisciplinary field called metagenomics (Imhoff 2016). It involves the isolation of the genomic DNA from environmental samples and after cloning and expressing in a culturable organism for further analysis (Handelsman 2004). The term metagenomics was coined to denote the meta-examination of the relatively similar microbial population residing in the various niche (Neelakanta and Sultana 2013). The metagenomic field was brought in the spotlight by the studies of DeLongs and his colleagues after they generated the metagenomic library of prokaryotes from sea-water. The 16s rRNA sequencing confirmed that the library belonged to the archaeon and was not cultured until that time (Stein et al. 1996).

The sequencing analysis has a unique role in the identification and functional annotation of metagenomic samples (Österlund et al. 2017). The very first metagenomic analysis coupled with shotgun sequencing was carried out to analyse viral, microbial diversity residing on the surface of sea-water, which has identified more than 65% of the sequences having no prior knowledge (Osunmakinde et al. 2018). Majority of the metagenome analysis has been performed concerning marine samples because two-thirds of the earth is occupied by water. Secondly, it provides the niche to diverse microbial communities which regulate the ecosystem (Aguiar-Pulido et al. 2016). Apart from this, the marine microbial population serves as a hidden treasure trove for novel pharmaceutical and industrially relevant metabolites (Hug et al. 2018). Progressively, the combination of metagenomics with High-throughput sequencing (HTS) technologies has opened the floodgates to find novel metabolites from microbes having clinical implications. Together these approaches have identified various important metabolites like biomass-degenerating enzymes from cow rumen, recognizing novel CRISPR (Clustered Regularly Interspaced Short Palindromic Repeats) systems and set up of gene index of the human microbiome (Seshadri et al. 2018; Stewart et al. 2018).

8.2 HTS Technologies for Metagenome Examination

HTS technology was introduced firstly in 2005, and since then, it is being improvised efficiently to provide greater accuracy and precision in terms of genome analysis (Reuter et al. 2015). The advancement in the HTS has led the foundation of extensive exploration of microbial communities by producing high throughput genomic data with rate and time effectiveness (Zhou et al. 2015). The array of HTS technology involves amplicon sequencing, whole-genome sequencing and shotgun metagenome sequencing (Petrosino et al. 2009). The pros of HTS over old-style sequencing convention involve its high-throughput data generation, absence of cloning and fewer tariffs (Ari and Arikan 2016). The critical step in this technology is to draw statistically significant inferences out of the generated data. In the following discussion, various HTS platforms which can be implemented to analyse metagenomes are highlighted (and summarized in Fig. 8.1).

8.2.1 Roche 454 Sequencer

Variants GS20, GS-FLX, GS-FLX Titanium and GS-FLX Titanium+.

Several bioactive metabolites have been obtained from metagenomic samples with immense therapeutic potential. Some examples include malacidin, fluoroquinolone, minimide and erdacin. GS20 was the first HTS variant introduced in 2005. This sequencer implements sequencing by synthesis approach in a picotitre plate, which gives 20 megabases of output in a single run and mean read size of 100 base pair (Pareek et al. 2011). The sequencer works on the mechanism of pyrosequencing, which involves NTP (Nucleotide Triphosphates) and nucleotide addition complementary to the sequencing strand is detected by the liberation of pyrophosphate (Harrington et al. 2013). Its high-end variant GS-FLX Titanium + generates around 850 megabases in a single run with a mean read size of 700–750 base pairs. The system was most suitable for the 16S rRNA sequencing as it can pave the highly capricious fragments of 16S rRNA. The GS-FLX variants were discontinued since December 2016 due to their cost, error rate and quantity of sample DNA required

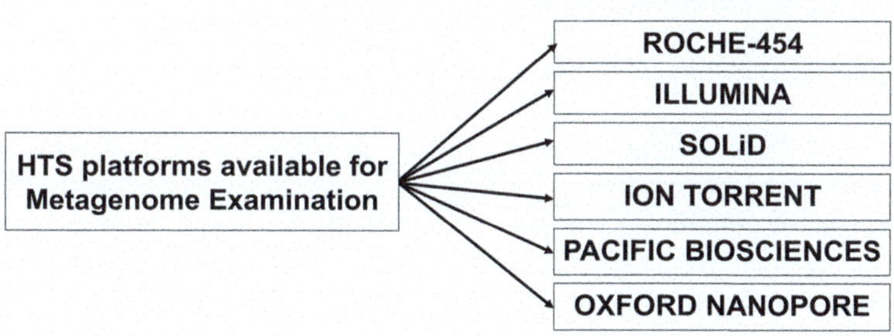

Fig. 8.1 Various HTS platforms available for metagenome examination

was higher in comparison with other HTS platforms, but it has produced an enormous amount of data which is not yet available in the scientific knowledge (Liu et al. 2012).

8.2.2 Illumina Sequencer

Variants GA I, II, HiSeq, MiSeq, NextSeq 500, HiSeq2500 and HiSeq X Ten.

The Illumina sequencer was introduced back in 2006 and got widely used by the scientists because of its lower cost. However, its major downside was smaller read size in its initial variants which were taken care of in the advanced variants. The MiSeq variant gives 2 × 300 base pairs of read-length (Quail et al. 2008). The above-discussed peculiarities made the scientists switch from the Roche 454 platform to Illumina sequencer. It utilises sequencing by synthesis tactic by the termination. The HiSeq 2500 variant produces optimum four billion pieces of 125 bases size for a single read in a paired-end manner dye (Schirmer et al. 2016). Its recent advancement HiSeq X Ten includes, as the name suggests coupling of ten HiSeq machines to obtain high-throughput data (Levy and Myers 2016). Illumina has also introduced the first small-sized sequencer regarded as NextSeq 500 (Buermans and den Dunnen 2014).

8.2.3 SOLiD (Sequencing by Oligonucleotide Ligation and Detection)

SOLiD was announced in 2006, by Applied Biosystems as a sequencing platform which utilises sequencing by ligation methodology (Pareek et al. 2011). The variant 5500 xl generates 300 gigabytes of data, 3 billion base pairs in a single run with the reading size of 75 base pairs. Despite abundant data generation and less rate for single-base sequencing, the comparatively smaller read size and cost for a single run were its major setbacks (Goodwin et al. 2016).

8.2.4 Ion Torrent Sequencer

Variants PGM (Personal Genome Machine, Proton and S5).

Ion torrent was the first organisation which has introduced small-scale sequencers in the form of PGM to the researchers in 2010. As a result, it received positive feedback and become a hotspot among researchers to perform sequencing analysis in comparatively lesser spending. The sequencing was carried out in a microtiter plate in which DNA stretches are incorporated to the beads when the DNA is supplemented to the sequencing strand it liberates the proton which results in a change of pH and sensed by the detector. Ion Proton, the high-throughput variant of the Ion torrent, generates ten gigabases of data with almost 50 million reads in a single run having read size of 200 base pairs (Lahens et al. 2017). The most significant advantage of PGM is that it can generate a large read size of about 400 base pairs (Henson et al. 2012). Their latest variant is Ion S5 which can generate 15 gigabytes of output with 60 to 80 million reads in a single run of size around 200 base pairs (Mehrotra et al. 2017).

8.2.5 Pacific Biosciences

Variant Pac Bio RS ll.

In 2012, Pacific biosciences launched its SMRT (Single Molecule Real-Time) Sequencing platform which performs sequencing by synthesis. Helicos biosciences were the first single-molecule sequencing platform, but the PacBio was able to obtain a distinguished reputation in SMRT sequencing area. It utilises CCS (Circular Consensus Sequencing) in order to obtain error-proof sequence stretches. The variant of the respective company performs sequencing by ZMW (Zero-Mode Waveguide) in which DNA polymerase is ligated to a unit DNA molecule. Subsequently, the incorporation of the DNA in the strand is recorded by detection of luminescence. Each of the four types of nucleotide are labelled with different fluorescent dyes which on incorporation to the synthesizing strand liberates different fluorescence. It generates 10,000 to 60,000 base pairs read with comparative precision of around 99.999 percent. Due to its exceptional annotation efficiency, they are regarded as best for shotgun metagenome analysis (Ardui et al. 2018).

8.2.6 Oxford Nanopore Sequencer

Variants MinIon, PromethION, SmidgION and VolTRAX.

The nanopore sequencer from Oxford utilises the state-of-the-art strand sequencing which can sequence the entire DNA fragment by detecting the change in electric current when passed through minute nanopores made of proteins. MinIon mk1B is a compact sequencer which can get coupled with any sort of computer for immediate data analysis. The PromethION sequencer offers 144,000 (3000 nanopore channels of 48 flow cells) channels for the sequencing purposes. Their SmidgION variants can get operated through a smartphone for instant analysis. The VolTRAX variant can be controlled through the Universal Serial Bus (USB) after sample load. The comparatively larger read size removes the necessity of shot-gun sequencing and thus bringing a revolt in the respective (Wanunu 2012).

8.3 Elucidation of Metagenomic Data

The analysis of metagenome data is planned explicitly to process concoction of genomes and contigs of different sizes. The elucidation of metagenome data involves the following stages.

8.3.1 Processing of Poor-Quality Reads

The poor-quality reads are initially processed by the utilities supporting the variant of sequencer used for the sequencing. One such utility is the FASTX-Toolkit (command-line based utility for pre-processing of FASTA/FASTQ data), apart from this, FastQC (quality check utility to process raw HTS data) is also implemented for the same purpose which also gives the overall figures of the FASTQ data. Tools such as Galaxy (multivariate genome analysis platform), SolexaQA (to view a graphical representation of sequence quality) and Lucy 2 (command-line based sequence cleaner and visualiser) are implemented to process FASTQ data. These tools utilise Q quality or Phred scores (measures the sequencing quality), whose verge relies on the variant of sequencer implemented.

8.3.2 Masking of Low-Complexity Reads

This is carried out with the help of utilities like DUST. After this, the reads/sequences which are sharing more than 95% identity are eliminated. Some tools like MG-RAST (MetaGenomic Rapid Annotation using Subsystems Technology) allows the user to eliminate reads which are almost matched with the genome of model organisms like human, fly, cow and mouse. The process is mediated by the Bowtie 2 (fast and efficient tool for the alignment of sequencing reads against reference sequence) utility.

8.3.3 Gene Identification

In this step "gene calling" is brought into action that allows the user to recognise genes which are present in reads/contigs. CDS (Coding DNA Sequence), non-coding RNA genes and some tools also allow the user to recognise CRISPR (Clustered Regularly Interspaced Short Palindromic Repeats). Metagene, FragGeneScan, Prodigal, MetaGeneMark and Orphelia helps to recognise the CDS genes by implementing ab-initio gene prediction. These utilities implement codon information to recognise regions of reads/contigs as introns and exons. They can be trained by using the user-oriented datasets. FragGeneScan is used for recognising prokaryotic genes and implemented by IMG/M (Integrated Microbial Genomes with Microbiome samples; a platform to perform metagenome comparative data analysis), EBI (European Bioinformatics Institute) Metagenomics and MG-RAST

(Metagenomics-Rapid Annotation using Subsystems Technology; a platform to carry out evolutionary and functional analysis of metagenomes). The prediction accuracy of FragGeneScan ranges from 65 to 70 percent. Non-coding RNA is predicted by the tools like tRNAscanto anticipate the tRNAs, on the other hand, rRNA genes are anticipated by personalised rRNA models for IMG (Integrated Microbial Genome)/MER (Microbiomes Expert Review) and MG-RAST implement homology-based search with SILVA (specialised database of aligned ribosomal RNA sequences of Eukarya, archaea and bacteria), RDP (Ribosomal Database Project). PILER-CR and CRT (CRISPR Recognition Tool) are implemented to anticipate CRISPR stretches.

8.3.4 Gene Annotation

The next step of the metagenomic data elucidation includes allocation of the functions to the genes. The objective is accomplished by the similarity—search method in which investigational sequence is compared with the database sequence having annotated genes information. The basic steps involved in metagenome annotation are shown in Fig. 8.2. The bigger size of the metagenomic data has made this process automated and computationally expensive. BLAST (Basic Local Alignment Search Tool) utility is implemented in high-end computing servers. The concept of multithread is implemented in which a process is separated into numerous CPU (Central Processing Unit)/GPU (Graphical Processing Unit) in order to obtain

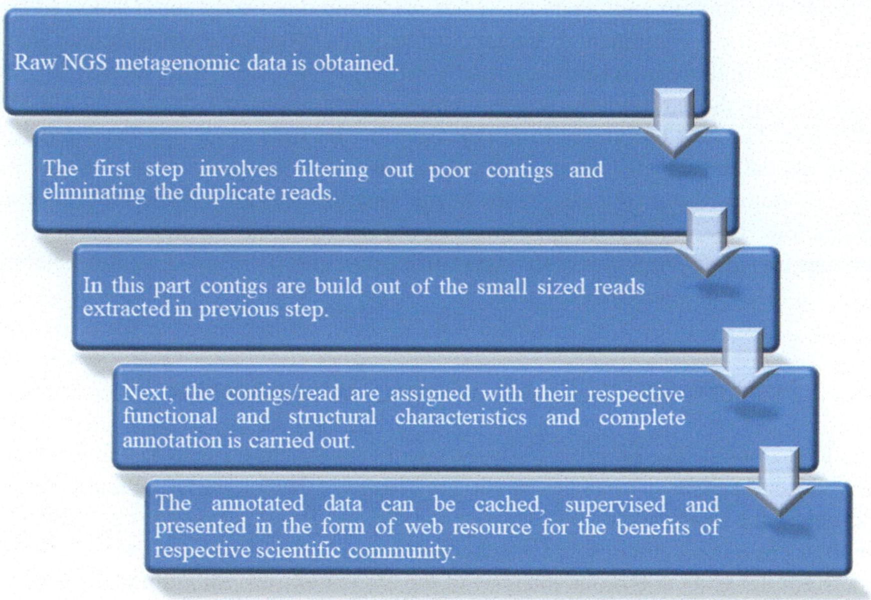

Fig. 8.2 General work-flow of the HTS metagenome annotation

the results in short time-span. The metagenomic data is annotated with the help of various databases like KEGG (Kyoto Encyclopedia of Genes and Genomes), egg-NOG (evolutionary genealogy of genes: Non-supervised Orthologous Groups), COG (Clusters of Orthologous Groups)/KOG (EuKaryotic Orthologous Groups) and protein databases like PFAM (Protein FAMily), Interpro and TIGRFAM (The Institute of Genomic Research's database of protein Families). The use of numerous databases mentioned above is brought into action for metagenomic data annotation.

IMG/MER uses Hidden Markov Model (HMM) profile to link the query set genes with the PFAM after which with the help of COG ortholog clustering is carried out. The PSSM (Position-specific Scoring Matrix) dataset is retrieved from the NCBI (National Center for Biotechnology Information) for functional assignment of proteins. On the other hand, genes are identified with the help of KEGG, and EC (Enzyme Commission) numbers and evolutionary analysis of the metagenome data is carried out by homology exploration.

IMG/MER contains a huge amount of genomic data which it utilises to retrieve extra annotation information. The first step in its workflow is the anticipation of the genes out of the metagenome and subsequently utilises other options to annotate those genes further. This leads to the recognition of PFAM, which is not determined in case of MG-RAST and results in comprehensive annotation parallel to the COG, which is the only protein family identification resource utilised by MG-RAST. One tailback for IMG/MER is the rapid increase in the gene counts which is not the case with the MG-RAST but because the query metagenome as in case of IMG/MER is subjected to PFAM analysis results in extensive annotation and reporting of the metagenome.

MG-RAST initially anticipate the genes present in the metagenome followed by searching for the homologs of those anticipated genes in the separated genomes. The process is carried out by a utility called BLAT (BLAST—Like Alignment Tool). It considers only those homologs whose identity is more than 70% thus omits considerable hits. The best homologs from the separated genome are further subjected to annotation rather than the metagenome. This turns into a drawback as the annotation is carried out on the substituted genes of the separated genome while ignoring the metagenome. Nevertheless, the plus point of implementing this method is the shorter time-pan for complete annotation. Apart from this, the database does not enlarge while the IMG/MER enlarge its mass.

8.4 Pharmaceutical Products from Metagenomes

In order to obtain the pharmaceutically significant metabolites, it is of immense importance to primarily considering the functionality. The objective of the research is to obtain the pharmaceutically validated active metabolite having preferred functionality. The advancement in HTS technologies has shifted the interest of researches to perform the sequence-based analysis of the clone libraries. Shotgun metagenome sequence analysis has resulted in not-only in a prompt classification of the biosynthetic gene clusters but also the anticipation of corresponding biochemical assembly.

However, the standalone bioinformatics and HTS approach can anticipate a limited number of gene clusters, but the improved facilitation has allowed the researchers to identify novel pharmaceutically significant active metabolites.

Nowadays, instead of functionally annotating the metagenome gene clusters, the research has shifted to the targeted screening, which considers the background of the metagenome under examination. In the upcoming part, we highlight various pharmaceutically active metabolites obtained through metagenome examination. Irrespective of the pipeline followed for the functional and structural characterisation, and the metagenome is a reservoir for several metabolite-synthesizing genes.

8.4.1 ET-743 (Yondelis)

In 1969, the examination of the marine squirt *Ecteinascidia turbinata* resulted in the identification of its anti-cancer properties and the structure of Ecteinascidin (ET-743) was elucidated in 1984, and presently it is a pharmaceutically validated anti-cancer metabolite. The practices to grow sea squirt in order to fulfil the pharmaceutical hunger was not that successful but alternatively, extensive artificial approaches were adopted to meet the pharma needs. The identification of ET-743 homology with bacteria-derived metabolites namely saframycin A (*Streptomyces lavendule*), safracin B (*Pseudomonas fluorescens*), saframycin Mx1 (*Myxococcus xanthus*) has resulted in the understanding that symbiotic bacterial communities synthesized ET-743. The metagenome sequence analysis of tunicate depicts that it regulates non-ribosomal peptide synthetase pathways by the expression of 25 genes. The extensive sequence annotation workflow has identified that the bioactive molecule is generated by *Candidatus Endoecteinascidia frumentensis*. The whole-genome size of the particular organism was identified approximately 631 kb. The determination of the associated pathway for the metabolite synthesis open the gates for the pharma industries to synthesize the respective metabolite along with its analogues at massive scale.

8.4.2 Bryostatins

In 1968, Bryostatin was found in *Bugula neritina*, which further caught the limelight because of its toxic action for the cancerous cells, explicitly targeting Protein Kinase C. The activity of the Bryostatin was estimated by more than 80 clinical, trials and the medication is used for the Alzheimer. Initially, it was found that Bryostatin was expressed by symbiotic relationships as numerous forms of the compound exists. Later, the cosmid library respective of B. *neritina* was constructed, and numerous corresponding clones were sequenced, which leads to the identification of 65 kb *bry*gene group. Further, the hybridisation studies were carried out on two E. *sertula* strains from a different host. In one strain, the genes were found adjoining while the other strain was having the respective gene cluster fragmented from the auxiliary genes. As the E. *sertula* is not-culturable, to meet the needs of the pharma industry, the *bry*gene cluster can be expressed in various host organisms.

8.4.3 Psymberin

Psymberin is a kind of polyketide with cell-toxicity activity against the tumour cells. It was obtained from the numerous sea sponges. The compound is of little importance because of its intricate structure, bioactivity and structure of the respective compound were elucidated in 11 years utilising more than 600 samples. The biosynthesis process of the compound was determined in the metagenome of *Psammociniaaff. Bulbosa*. Sample of *Psammociniaaff. Bulbosa* were collected from scuba diving at Milne Bay, New Guinea. Further, the protein sequence-based alignments of psymberin and other related groups were generated. Amplicons were generated using the primer-based amplification approach. Total sponge DNA was isolated and two libraries (3,20,000 and 9,00,000 clones) respective of *Psammociniaaff. Bulbosa* were generated following PCR based screening using psymEAD-Yyspez2-forward and psymEAD-Yyspez1-reverse primers to obtain the PKS gene clusters. (Haas 2009). The genomic composition analysis of the respective compound suggests its derivation from the bacteria.

8.4.4 Onnamides

The tumour targeting particularity of the mycalamide and pederin inhibits the replication and translation mechanism even at the slight concentration of 1 ng/ml. The clinical trial study suggests that the implication of the respective compounds lead to increase the life-span of cancer-induced mice. The assistance of metagenome analysis has found that these compounds are derived from non-culturable *Pseudomonas* linked with *Paederusfuscipes*. The homology study of the pederin results in the identification of structurally and functionally similar compounds in *Lithistida*. The pederin-led analysis of polyketide synthase used as amplification template was obtained from *Theonella swinhoei* metagenome and subsequently determined the biosynthesis pathway for onnamide. Advanced analysis of the T. *swinhoei* metagenome depicts that these polyketide synthases can only be derived from those sponges which had comprises pederin homologs formerly. Finally, it was known that onnamides are derived from non-culturable *Candidatus Entotheonella spp*.

8.4.5 Patellazoles

Patellazoles were extracted from the tunicate during 1980, and they have gained importance due to their pharmacological significance. The respective metabolite was able to show potential antifungal action and cell-toxicity action to human cell-lines.

The chief symbiont for this metabolite was *L. patella* and *C. albicans*. The structural units of the patellazole contain acetate and thiazole ring has led to the assumption that it may be synthesized through polyketide synthase and non-ribosomal peptide synthetase pathway respectively. The assumption was tested by carrying out sequence-based metagenome analysis of tunic-cloaca niche and gave-off negative findings. The PCR (Polymerase Chain Reaction) analysis revealed the patellazole synthesis mediated through trans-acyltransferase family from the miniature zooids. Shotgun sequence analysis of isolated zooid DNA reveals the 86 kb genome corresponding to trans-acyltransferase polyketide synthase pathways. The genome was contemplated to possess by *Candidatus Endolissoclinum faulkneri*.

8.4.6 Calyculin A

The compound was extracted from sea sponge *Discodermia calyx* in 1986 which possess high cellular-toxicity abilities. The biosynthesis of the respective compound is mediated by a combination of non-ribosomal peptide and polyketide pathways, respectively. The homologs respective of Calyculin were also identified, which suggests its derivation regulated by symbiotic interaction. The biosynthetic gene cluster for the Calyculin was determined with the aid of metagenome analysis. Metagenome examination of *D. calyx* reveals 150 kb of the gene cluster. Taking this gene cluster as a template with the help of molecular biology analysis, the Calyculin synthesis pathway was found to be possessed by filamentous bacteria. Further, 16s rRNA sequence analysis of the respective bacteria showed 97% homology with *Candidatus Entotheonella* factor obtained from *T. swinhoei* sponge.

8.4.7 Polytheonamides

This group of compounds possess cellular-toxicity and are extracted from *T. swinhoei* sponges. The compound is synthesized through non-ribosomal peptide synthetase, and the peptide length of the respective compound is 48 amino acids. The PCR analysis of *T. swinhoei* metagenome revealed that the compound is derived from the ribosome and is a product of the symbiotic relationship of the bacteria. Subsequent analysis revealed that the producer of polytheonamide was non-culturable *Entotheonella* spp. This species was responsible for the production of polytheonamide and onnamide metabolites.

8.5 Conclusion

Metagenomic examination of the samples from the different environmental condition of niches has enabled the researchers to dive into the oceans of screening a large number of pharmaceutically relevant active metabolites. The metagenome examination with the help of HTS approaches has enabled the researchers to not only retrieve the metagenome data but also perform the annotation effectively. We have discussed various HTS platforms which can be utilised for the metagenome examination along with their pros and cons. The strategies which can be employed to perform the metagenome analysis has also been discussed along with some useful resources. The main objective of the book chapter was to draw the attention of researchers towards the HTS usefulness in metagenome examination so that this exponentially growing field not only receive the appreciation but also direct the intellect of wet-lab researchers towards designing their work-flow in a manner by which the distinguished properties of both wet lab and dry lab analysis can be utilised to serve the human society at their best.

References

Aguiar-Pulido V, Huang W, Suarez-Ulloa V et al (2016) Metagenomics, metatranscriptomics, and metabolomics approaches for microbiome analysis. Evol Bioinforma 12:5–16

Ardui S, Ameur A, Vermeesch JR, Hestand MS (2018) Single molecule real-time (SMRT) sequencing comes of age: applications and utilities for medical diagnostics. Nucleic Acids Res 46:2159–2168

Ari Ş, Arikan M (2016) Next-generation sequencing: advantages, disadvantages, and future. In: Hakeem KR, Tombuloğlu H, Tombuloğlu G (eds) Plant omics: trends and applications. Springer, Berlin, pp 109–135

Arnold JW, Roach J, Azcarate-Peril MA (2016) Emerging technologies for gut microbiome research. Trends Microbiol 24:887–901

Buermans HPJ, den Dunnen JT (2014) Next generation sequencing technology: advances and applications. Biochim Biophys Acta Mol basis Dis 1842:1932–1941

Goodwin S, McPherson JD, McCombie WR (2016) Coming of age: ten years of next-generation sequencing technologies. Nat Rev Genet 17:333–351. https://doi.org/10.1038/nrg.2016.49

Haas MJ (2009) Polyketide pas de deux. Sci Exch 2:898–898. https://doi.org/10.1038/scibx.2009.898

Handelsman J (2004) Metagenomics: application of genomics to uncultured microorganisms. Microbiol Mol Biol Rev 68:669–685. https://doi.org/10.1128/MMBR.68.4.669-685.2004

Harrington CT, Lin EI, Olson MT, Eshleman JR (2013) Fundamentals of pyrosequencing. Arch Pathol Lab Med 137:1296–1303. https://doi.org/10.5858/arpa.2012-0463-RA

Henson J, Tischler G, Ning Z (2012) Next-generation sequencing and large genome assemblies. Pharmacogenomics 13:901–915

Hug JJ, Bader CD, Remškar M et al (2018) Concepts and methods to access novel antibiotics from actinomycetes. Antibiotics 7:44

Imhoff J (2016) New dimensions in microbial ecology—functional genes in studies to unravel the biodiversity and role of functional microbial groups in the environment. Microorganisms 4:19. https://doi.org/10.3390/microorganisms4020019

Lahens NF, Ricciotti E, Smirnova O et al (2017) A comparison of Illumina and Ion Torrent sequencing platforms in the context of differential gene expression. BMC Genomics 18:602. https://doi.org/10.1186/s12864-017-4011-0

Levy SE, Myers RM (2016) Advancements in next-generation sequencing. Annu Rev Genomics Hum Genet 17:95–115. https://doi.org/10.1146/annurev-genom-083115-022413

Liu L, Li Y, Li S et al (2012) Comparison of next-generation sequencing systems. J Biomed Biotechnol 2012:1–11. https://doi.org/10.1155/2012/251364

Mehrotra M, Duose DY, Singh RR et al (2017) Versatile ion S5XL sequencer for targeted next generation sequencing of solid tumors in a clinical laboratory. PLoS One 12:e0181968. https://doi.org/10.1371/journal.pone.0181968

Neelakanta G, Sultana H (2013) The use of metagenomic approaches to analyze changes in microbial communities. Microbiol Insights 6:MBI.S10819. https://doi.org/10.4137/mbi.s10819

Österlund T, Jonsson V, Kristiansson E (2017) HirBin: high-resolution identification of differentially abundant functions in metagenomes. BMC Genomics 18:1–11. https://doi.org/10.1186/s12864-017-3686-6

Osunmakinde CO, Selvarajan R, Sibanda T et al (2018) Overview of trends in the application of metagenomic techniques in the analysis of human enteric viral diversity in Africa's environmental regimes. Viruses 10:429

Pareek CS, Smoczynski R, Tretyn A (2011) Sequencing technologies and genome sequencing. J Appl Genet 52:413–435

Petrosino JF, Highlander S, Luna RA et al (2009) Metagenomic pyrosequencing and microbial identification. Clin Chem 55:856–866

Quail MA, Kozarewa I, Smith F et al (2008) A large genome center's improvements to the Illumina sequencing system. Nat Methods 5:1005–1010. https://doi.org/10.1038/nmeth.1270

Reuter JA, Spacek DV, Snyder MP (2015) High-throughput sequencing technologies. Mol Cell 58:586–597

Schirmer M, D'Amore R, Ijaz UZ et al (2016) Illumina error profiles: resolving fine-scale variation in metagenomic sequencing data. BMC Bioinf 17:125. https://doi.org/10.1186/s12859-016-0976-y

Seshadri R, Leahy SC, Attwood GT et al (2018) Cultivation and sequencing of rumen microbiome members from the Hungate1000 Collection. Nat Biotechnol 36:359–367. https://doi.org/10.1038/nbt.4110

Stein JL, Marsh TL, Wu KY et al (1996) Characterization of uncultivated prokaryotes: isolation and analysis of a 40-kilobase-pair genome fragment from a planktonic marine archaeon. J Bacteriol 178:591–599. https://doi.org/10.1128/jb.178.3.591-599.1996

Stewart EJ (2012) Growing unculturable bacteria. J Bacteriol 194:4151–4160

Stewart RD, Auffret MD, Warr A et al (2018) Assembly of 913 microbial genomes from metagenomic sequencing of the cow rumen. Nat Commun 9:870. https://doi.org/10.1038/s41467-018-03317-6

Wanunu M (2012) Nanopores: a journey towards DNA sequencing. Phys Life Rev 9:125–158

Zhou J, He Z, Yang Y et al (2015) High-throughput metagenomic technologies for complex microbial community analysis: open and closed formats. MBio 6:e02288-14. https://doi.org/10.1128/mBio.02288-14

Epidemiological Perspectives of Human Health Through Metagenomic Research

9

Hemender Singh, Indu Sharma, and Varun Sharma

Abstract

Metagenomics is an approach to analyzing environmental nucleic acids. With human health, metagenomics has played a significant role in understanding the human microbiome (gut, skin, vaginal, airway microbiome, among others) to a greater extent. This gave a new insight to view and understand the role of human microbiota with the disease pathology. Traditional diagnostic approaches were dependent on culture-based techniques which require prior knowledge about that microbe. However, with the advent of high throughput techniques and metagenomic techniques like high-throughput sequencing, it becomes easier to identify novel pathogen *de-novo* and also helps in determining the functional and taxonomic categorization. Metagenomics also aid in the identification of the antibiotic-resistant phenotypes and virulence genes that helps in clinical diagnosis and outbreak investigations.

Keywords

Cancer · Epidemiological · Metagenomic · Microbiome

9.1 Introduction

The relationship of human health and pathogens have been known for long, but with the advent of metagenomics and advances in the techniques which are culture-independent has allowed further studies on the human health (Tyson et al. 2004). The association of gut microbiome with host health has allowed the understanding of

H. Singh
School of Biotechnology, Shri Mata Vaishno Devi University, Katra, India

I. Sharma · V. Sharma (✉)
Ancient DNA Laboratory, Birbal Sahni Institute of Palaeosciences, Lucknow, Uttar Pradesh, India

R. S. Chopra et al. (eds.), *Metagenomics: Techniques, Applications, Challenges and Opportunities*, https://doi.org/10.1007/978-981-15-6529-8_9

how environment and lifestyle along with population of the gut flora affect the overall functioning and health of the host (Kelsen and Wu 2012). Since the introduction of metagenomics around two decades ago, this technique has changed the study of the microbiota. Hence, the scientific community relates the microbial flora within the human body to its health.

Microbiota can be assumed as a multi-celled organ which communicates with other organs for well-functioning of the body (Cani and Knauf 2016). However, considering the factors such as a large number of habitats within a host, interactions of the species with the host, site of localization and external environment with which the host interacts, the microbiota can also be described as a dynamic ecological environment.

The gastrointestinal gut flora in humans have now reached a mass of around 2 kg, and the genomes it contains codes for probably plenty of different functions that our cells do not undertake (Wen and Duffy 2017). The increasing number of scientific claims that we are super-beings with a microbial load 10x higher than human cells which should be taken into account as part of human physiology (Gill et al. 2006; Floch 2011).

9.2 Metagenomics in Human Microbiome Project

The microbiome can be described as genome collected from all the microorganisms present within the environment or in the case of humans, the body (Kelsen and Wu 2012). Microbiome should not be confused with microbiota. Though both are closely related, the microbiota is the term used when the microbial population is talked in reference to its localization within the environment.

To understand the microbial population and its diversity within and the human body, Human Microbiome Project was launched in 2008, having a funding of $157 million (Martín et al. 2014). The project developed research resources for studying microbial population and its interaction with the host. Samples from healthy volunteers were collected and studied. Four thousand seven hundred eighty-eight specimens in total were screened and phenotyped from 242 adults (https://hmpdacc.org/). Sites within or on the body where the microorganisms localize, starting from the skin, nasal and buccal mucosa, mammary glands, placenta, uterus, ovarian follicles, seminal fluids, the gastrointestinal tract where an abundance of microorganisms can be found were the sample collection sites. The oral and stool sample contained the most diverse communities, while the vaginal communities were rather simple (Turnbaugh et al. 2007). In HMP, no one taxa were observed to be present within or on the host universally, at the sequencing depth employed. Although fewer dominant taxa were observed as well, and it was noticed that they were personalized highly and varied among individuals. The ecology of microbial metabolism and functional pathways in the microbiome communities were assessed as well. It was observed that the communities were much more functional and constant in relative abundances of pathways than were organism abundances. This confirmed that the entire human microbiome has an ecological property. Pathways were omnipresent among organism and body habitats. Ribosome and translational machinery; nucleotide charging and ATP synthesis, and glycolysis being the most abundant pathways and reflecting the basics of host-associated microbial life (Huttenhower et al. 2012). It was observed that the host characteristics such as age, gender, body temperature, body mass index (BMI), among others are associated with clades and metabolism in the microbiota (Huttenhower et al. 2012).

The HMP through the extensive sampling provided a basic characterization of an average healthy western population making it possible to relate the microbiome-based disorders with possible diseases. Utilizing both 16S and metagenomic profiling from an organism and functional data, along with characterization, allowed studies to move have a broader perspective towards microbiome and microbial communities. These various communities present within are related directly to the health of the host, and so understanding their existence proved to be of enormous importance. As per the data available in different pieces of literature and databases, the human microbiome data of Indian population is scanty. Keeping in light the human population heterogeneity of Indians, it is a need of the hour to have such a database to have an insight into the microbiome of the Indian population. It may aid in the diagnosis of the disease and outbreaks investigations which is prevalent in Indians.

With the advancement in technology, various ecosystems present within the host body has been characterized. These ecosystems are of complex communities. Gastrointestinal tract (GIT) and vaginal ecosystems are the most studied and well characterized. Figure 9.1 illustrates the metagenomic workflow for unraveling the microbial diversity of the skin, gastrointestinal tract, vagina and the airways. The Dysbiosis is commonly linked to GIT, but it can occur on exposed surfaces and mucus membranes such as skin, vagina and respiratory tract. There is a significant impact of microbial variations on the host health. Research teams and scientists around the world are trying to associate the microbiome and human health and identify potential correlations by studying different disease conditions and microbiome.

9.3 The Microbiome of Gastrointestinal Tract (GIT)

The human intestine is not as rich as soil and water in terms of serving as a habitat for diverse microorganism and so can be described as physico-chemically harsh as compared to the natural reservoirs of the bacterial population. The human gut microbiome is considered beneficial for the host body as it has a crucial role in stimulation and maturation of the immune system, resistance against pathogen attack, among others. Along with the bacterial population in the intestine, a minority population of other microorganism is present as well, such as yeast, fungi, viruses, archaea (Belkaid and Hand 2014).

Culture-independent techniques have allowed characterization and study of an organism at a molecular level and changed the perspective of research in fields where cultural growth was a limitation. 16S rDNA-based approaches (gene sequencing)

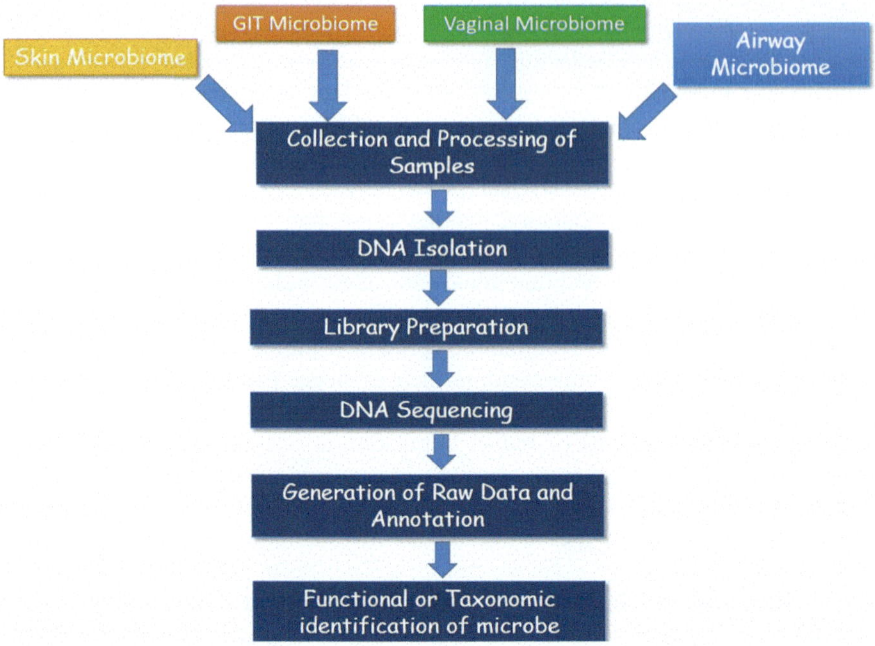

Fig. 9.1 A simplified metagenomic approach for identification and taxonomic categorization of microbes in the human body microbiome using high throughput sequencing

and other development metagenomics approach has changed the understanding of gut microbiome (Jovel et al. 2016). The reference gene catalogue, European project MetaHIT and the American Human Microbiome Project have been the source for reviews on the metagenomic exploration of the human intestinal microbiome (Blanco-Míguez et al. 2017).

The MetaHIT focused its efforts on defining core gut microbiome, while also reporting 3.3 million non-redundant genes in the human gut microbiome alone. The average human's intestinal microbiome in now well-defined and known to be diverse with firmicutes and Bacteroidetes phyla, both dominating and representing up to 90% of the microbiome population (Lloyd-Price et al. 2016). One concept is 'enterotype' which includes three robust clusters mainly, *Bacteroids, Prevotella and Ruminococcus,* and abundance and proportions of these enterotypes are associated with long-term dietary habits and vary among individuals (Voreades et al. 2014). The enterotypes are primarily affected by factors of age, geographical region nutritional details, among others. Furthermore, these factors have an impact on their composition. Thus, this concept can help simplify the complex microbiome known to us.

The parallel evolution of GIT and microbiome has led to the adaption of favorable habitation in the host organism (human) (Foster et al. 2017). The microbiome in the intestine has 150 times higher genes than the human genome, highlighting the significance of the equally-important genome of human physiology. The microbiota of human gut works as a barrier inhibiting the pathogen growth and its colonization (Kamada et al. 2013). High throughput sequencing of human biome has facilitated the efficient and unbiased characterization of a diverse microbial and viral continuum. This will also aid in the expansion of therapeutics based on human physiology.

Recent advancement in metabolomics has provided an insight into the structure and function of the human microbiome (Martín et al. 2014). The same approach can be used to analyze microbiome composition in a genetically susceptible host for prediction of chronic inflammation.

9.4 Microbiome of Vagina

The vaginal microbiome has a significant impact on human growth, physiology and immunity. This inhabitants of mutual bacteria constitute the host's first line of defense by removing non-native pathogenic microbes (Pickard et al. 2017). The first human vagina microbiology study identified lactobacilli as the commonly found microorganisms, accounting for more than 70% of all microorganisms isolated from healthy women's vaginal exudates (Martin 2012).

The vaginal ecosystem is complicated due to its physiological roles like menstrual cycle and personal practices like contraception among others, as a result of physiological and microbiological factors, it remains stable over the long term (Martin 2012).

Virginia Commonwealth University initiated The Vaginal Human Microbiome Project (VaHMP) intending to study the complex vaginal microbiome concerning different diseases and its unpredictable aftereffects in different physiological conditions (http://vmc.vcu.edu/). Various studies have established different vaginal microbiome profiles also known as "vagitypes", many of which are subjugated by a single bacterial species or family or class. Caution should be taken with this definition, however, because in these atypical microbiomes, it is not possible to rule out a transitional state between disease and health altogether. The differences in vaginal microbiome profiles were observed in populations with a difference in geographical origin and ethnicity, as well as during physiological changes in human (menstrual cycles, menopause, after and before and pregnancy) (Martin 2012). Owing to other environmental influences and sexual behaviors the vaginal microbiome often experiences shifts. Changes in the vaginal environment during pregnancy (estrogen and progesterone rates, epithelium thickness, and increased glycogen production) contribute to a substantial alteration in vaginal microbiota diversity which may be associated with pregnancy.

9.5 Microbiome of Skin

Skin is considered as a physical barrier for many infections, but it harbors a population of many beneficial symbiotic and commensal microbes, including bacteria viruses, fungi, among others (Martín et al. 2014). Due to the presence of such a complex ecosystem of microbes helps in preventing the pathogens from breaching this physical barrier by biological competition. With molecular biology, it becomes easier to identify diverse skin microbiota which was previously very tedious to do by using culture-based methods.

Introduction of metagenomics has led to the construction of libraries that helps in understanding the functions of the several taxa of bacteria *Staphylococcus*, *Propionibacterium*, *Corynebacterium*, and many more. It is interesting to know that this approach is not only limited to the identification of bacteria, but it has also helped to identify viruses such as human polyomavirus irrespective of the skin condition. This metagenomic approach will help in the identification of viral microbiome of the skin with different physiological conditions.

9.6 Microbiome of Airways

Airways of the human body constitute of the respiratory system from nose to lungs. This airway passage is a hub of many beneficial microbes and some pathogens (Hillman et al. 2017). In healthy individuals, the microbiome of air passage is homogeneous. It occurs in a gradient form with higher biomass gradient towards the upper tract with decreasing biomass gradient towards the lower tract.

Metagenomic studies on young children revealed that the microbiota of nasopharynx varies with seasons. In winters, it was found that infections of Proteobacteria and Fusobacteria were prominent, whereas, in springs, the infections of Bacteroidetes and Firmicutes were observed (Hanada et al. 2018). Talking of viral disease and their diagnosis, metagenomics led to the identification of both known and new viruses in the diagnosis of several nasopharyngeal diseases (Yozwiak et al. 2012).

Till now, very less is known about the lower microbiome of the respiratory tract even though it is known that various infectious agents were found to be associated with the pathology of the chronic lung diseases, but more research is needed to explore and validate the lower airway microbiome to uncover the pathogenesis of different lung conditions.

9.7 Human Microbiome and Cancer

As discussed earlier, the human microbiome is the collection of the genome of bacteria, fungi, viruses, protists, archaea found in and on the human body. This microbiome is found to be altered in many diseases. Talking about the involvement of metagenomics in therapeutics, type 2 diabetes is the first disease for which the association of microbiota with the disease was studied using metagenome-wide association studies (MWAS) (Wang and Jia 2016). Similarly, with the involvement of metagenomics and thus exploring human microbiome revealed that altered microbiota was found to be substantially affecting the cancer risk. Some studies indicate a positive association of some bacteria with specific cancer types. One such example is *Helicobacter pylori* with gastric cancers.

Following are some of the cancer types with known microbial cancer-causing microbial agents.

9.7.1 Esophageal and Gastric Cancer

One of the bacterially associated cancer types is gastric cancer which was found to be associated with the infection of *H. pylori* with over twice the increased risk of gastric cancer. According to the International Agency for Research on Cancer, 1994, *H. Pylori* was categorized as a class 1 carcinogen. *H. pylori* affect almost 50% of the global population and cause gastric inflammation, atrophy of epithelium and dysplasia because of the induction of strong immune response in a host.

However, in contrast to gastric cancer, *H. pylori* are associated with the lower risk of esophageal adenocarcinoma (Fox and Wang 2007). A study on 63 antral mucosal and 18 corpus mucosal samples where pyrosequencing was done, and they found a significant association of only *H. pylori* with gastric cancer was found (Jo et al. 2016).

A study on rat model revealed that high levels of *Streptococcus pneumoniae* were detected in normal epithelium and the Barrette's esophagus, in comparison to the tumor-adjacent normal epithelium, the reason may be that the bacteria first causes inflammation at the site and then leaves the cells when they are damaged and infect the surrounding tissues (Zaidi et al. 2016). Another metagenomic study on oral microbiome indicated the association of *Fusobacterium nucleatum*in esophageal cancer with shorter survival that might be indicating its role as a prognostic biomarker for esophageal cancer (Yamamura et al. 2016).

9.7.2 Lung Cancer

Lung cancer is one of the most dreadful cancer and a leading cause of cancer deaths worldwide with shocking facts of exceeding the number of deaths combined for the breast, prostate, kidney and colon cancer (Tsay et al. 2018). As discussed earlier in this chapter that the microbiome of the respiratory tract varies as we move from upper to lower respiratory tract considering this in view a study was carried out on 39 individuals with lung cancer and 36 individuals who were diagnosed as non-cancer. Also, ten more samples of healthy controls were included. 16S rRNA sequencing was performed to determine the lower airway microbiota. It was interesting to know that the samples were found to be enriched with the *Streptococcus* and *Viellonella,* which was further confirmed by in vitro models of airway epithelium (Tsay et al. 2018). These kinds of studies provide evidence that metagenomics has made a significant contribution to the field of medical diagnosis.

9.7.3 Other Cancer Types

With the help of metagenomics and high throughput sequencing techniques, it becomes easier to identify the microbiome or infectious agent associated with any cancer type. As in breast cancer, it was observed that subjects with breast cancer were to have increased abundance of *Corynebacterium*, *Staphylococcus*, Actinomyces and Propionibacteriaceae in urinary microbiome whereas decreased levels of *Methylobacterium* in breast tissue (Wang et al. 2017). Similarly, in another study on pancreatic cancer, the involvement of 16S rRNA sequencing to determine the oral microbiota between cases and controls were observed, and the results revealed that individuals carrying *Porphyromonas gingivalis* and *Aggregatibacter actinomycetemcomitans* were found to be associated with a high risk of pancreatic cancer. In contrast, individuals constituting the phylum *Fusobacteria* and its genus *Leptotrichia* were found to be associate with decreased risk of having pancreatic cancer (Fan et al. 2018). Also from the research on cervical cancer, it becomes clear that women with great vaginal microbiome diversity specifically consisting of *Lactobacillus gasseri* and *Gardnerella vaginalis* were more susceptible to Human papillomavirus (HPV) infection and remission rate in comparison to the women with high levels of *Atopobium* (Champer et al. 2018).

9.8 Metagenomics in Diagnostics

Starting our discussing on diagnostics, it becomes crucial to understand the disadvantages of the earlier diagnostic methods and need of metagenomics in the diagnostics. The traditional diagnostic methods were very cumbersome includes microscopy; culture-based methods, among others, and were based on prior knowledge. This makes it very difficult to identify new pathogens responsible for a disease. As far as metagenomics is concerned, it is non-culture based method and based on analyzing the environmental genome irrespective of its origin, methods like shotgun sequencing made it easier to identify and categorize new pathogens based on taxonomy and function.

This metagenomic sequencing data also helps in determining the antibiotic resistance genes in the bacteria, genes responsible for virulence that will help in informing the outbreak investigations (Miller et al. 2013).

It is interesting to know that metagenomics is not helping in the diagnosis of new diseases; this approach has uncovered the diagnosis of the infection that has been found in the cadavers. A study in 2011, on Tyrolean ice mummy Ötzi genome revealed that the sequences found were from the bacterium *Borrelia burgdorferi*, detecting the first known case of Lyme disease (Pallen 2014).

Only a few disadvantages are associated with metagenomics is the technical workforce and the technique is a bit expensive (Pallen 2014), also there is a chance of contamination as observed in the 2009 pandemic of H1N1 infection where one sample showed 97% sequence similarity with Ebola virus. However, after further investigation, it was found to be due to contamination (Miller et al. 2013). These problems will be diluted with time and with the advent of new technologies.

9.9 Conclusion

In this chapter, the role of metagenomics in the epidemiology of the disease is explained. Talking about human health, it becomes pertinent to understand the human microbiome. Since any dysbiosis may lead to the diseased condition. Metagenomics helped us in understanding the human microbiome from a different perspective. With the advancement in the high throughput sequencing techniques and metagenomics, many uncovered areas were explored such as the effect of the microbiome on the susceptibility of different cancers, the evolution of microbes, antibiotic resistance in bacteria and many more that ultimately helps in the diagnosis of the disease and outbreaks investigations. Recently metagenomics approach is used to explore the fungal microbiome (mycobiome) of the human body.

More research is needed in the field of metagenomics to overcome its disadvantages like contamination, cost of the technique that will be a boon in the field of human health and clinical diagnosis.

References

Belkaid Y, Hand TW (2014) Role of the microbiota in immunity and inflammation. Cell 157:121–141

Blanco-Míguez A, Gutiérrez-Jácome A, Fdez-Riverola F et al (2017) MAHMI database: a comprehensive MetaHit-based resource for the study of the mechanism of action of the human microbiota. Database 2017:baw157

Cani PD, Knauf C (2016) How gut microbes talk to organs: the role of endocrine and nervous routes. Mol Metab 5:743–752

Champer M, Wong AM, Champer J et al (2018) The role of the vaginal microbiome in gynaecological cancer. BJOG 125:309–315

Fan X, Alekseyenko AV, Wu J et al (2018) Human oral microbiome and prospective risk for pancreatic cancer: a population-based nested case-control study. Gut 67:120–127

Floch MH (2011) Intestinal microecology in health and wellness. J Clin Gastroenterol 45:S108–S110

Foster KR, Schluter J, Coyte KZ, Rakoff-Nahoum S (2017) The evolution of the host microbiome as an ecosystem on a leash. Nature 548:43–51

Fox JG, Wang TC (2007) Inflammation, atrophy, and gastric cancer. J Clin Invest 117:60–69

Gill SR, Pop M, DeBoy RT et al (2006) Metagenomic analysis of the human distal gut microbiome. Science 80(312):1355–1359

Hanada S, Pirzadeh M, Carver KY, Deng JC (2018) Respiratory viral infection-induced microbiome alterations and secondary bacterial pneumonia. Front Immunol 9:2640

Hillman ET, Lu H, Yao T, Nakatsu CH (2017) Microbial ecology along the gastrointestinal tract. Microbes Environ 32(4):300–313

Huttenhower C, Gevers D, Knight R et al (2012) Structure, function and diversity of the healthy human microbiome. Nature 486:207

Jo HJ, Kim J, Kim N et al (2016) Analysis of gastric microbiota by pyrosequencing: minor role of bacteria other than *Helicobacter pylori* in the gastric carcinogenesis. Helicobacter 21:364–374

Jovel J, Patterson J, Wang W et al (2016) Characterization of the gut microbiome using 16S or shotgun metagenomics. Front Microbiol 7:459

Kamada N, Chen GY, Inohara N, Núñez G (2013) Control of pathogens and pathobionts by the gut microbiota. Nat Immunol 14:685–690

Kelsen JR, Wu GD (2012) The gut microbiota, environment and diseases of modern society. Gut Microbes 3:374–382

Lloyd-Price J, Abu-Ali G, Huttenhower C (2016) The healthy human microbiome. Genome Med 8 (1):1–11

Martin DH (2012) The microbiota of the vagina and its influence on women's health and disease. Am J Med Sci 343:2–9

Martín R, Miquel S, Langella P, Bermúdez-Humarán LG (2014) The role of metagenomics in understanding the human microbiome in health and disease. Virulence 5:413–423

Miller RR, Montoya V, Gardy JL et al (2013) Metagenomics for pathogen detection in public health. Genome Med 5:81

Pallen MJ (2014) Diagnostic metagenomics: potential applications to bacterial, viral and parasitic infections. Parasitology 141:1856–1862

Pickard JM, Zeng MY, Caruso R, Núñez G (2017) Gut microbiota: role in pathogen colonization, immune responses, and inflammatory disease. Immunol Rev 279:70–89

Tsay J-CJ, Wu BG, Badri MH et al (2018) Airway microbiota is associated with upregulation of the PI3K pathway in lung cancer. Am J Respir Crit Care Med 198:1188–1198

Turnbaugh PJ, Ley RE, Hamady M et al (2007) The human microbiome project. Nature 449:804–810

Tyson GW, Chapman J, Hugenholtz P et al (2004) Community structure and metabolism through reconstruction of microbial genomes from the environment. Nature 428:37–43

Voreades N, Kozil A, Weir TL (2014) Diet and the development of the human intestinal microbiome. Front Microbiol 5:494

Wang H, Altemus J, Niazi F et al (2017) Breast tissue, oral and urinary microbiomes in breast cancer. Oncotarget 8:88122–88138

Wang J, Jia H (2016) Metagenome-wide association studies: fine-mining the microbiome. Nat Rev Microbiol 14:508–522

Wen L, Duffy A (2017) Factors influencing the gut microbiota, inflammation, and type 2 diabetes. J Nutr 147:1468S–1475S

Yamamura K, Baba Y, Nakagawa S et al (2016) Human microbiome *Fusobacterium nucleatum* in esophageal cancer tissue is associated with prognosis. Clin Cancer Res 22:5574–5581

Yozwiak NL, Skewes-Cox P, Stenglein MD et al (2012) Virus identification in unknown tropical febrile illness cases using deep sequencing. PLoS Negl Trop Dis 6:e1485

Zaidi AH, Kelly LA, Kreft RE et al (2016) Associations of microbiota and toll-like receptor signaling pathway in esophageal adenocarcinoma. BMC Cancer 16:52

Metagenomic Applications of Wastewater Treatment

10

Mamta Sharma and Neeta Raj Sharma

Abstract

Many parts of the world do not have adequate infrastructure to treat all the wastewater generated from various sources. Discharge of this untreated or poorly treated wastewater is a source of pathogens that may lead to numerous waterborne diseases. Surveillance of pathogens is essential for water safety and public health. Traditionally bacterial indicators and polymerase chain reaction-based methods have been employed for detection of pathogens from wastewater. Recent advancements in molecular biology involving genome sequencing, also called as Metagenomics are emerging as a powerful tool that allows detection and comprehensive analysis of microbes and viruses present in wastewater. Metagenomics, along with bioinformatics, is now being increasingly employed for characterization of viruses and discovering novel variants. Viruses are the most abundant entities in water, but most of them are not yet characterized, and their roles not yet elucidated. This chapter attempts to highlight the ability of bacteriophages to be used as alternative biocontrol agents in wastewater treatment. It also presents various other roles that phages can have in the wastewater treatment process. The chapter also discusses tradition methods used for detection of pathogens and the revolution that metagenomics has brought in this area. Finally, various applications of metagenomics in wastewater treatment are also discussed.

Keywords

Bioremediation · Metagenomics applications · Phage therapy · Viral metagenomics

M. Sharma (✉) · N. R. Sharma
School of Bioengineering and Biosciences, Lovely Professional University, Phagwara, Punjab, India
e-mail: mamta.18431@lpu.co.in; neeta.raj@lpu.co.in

R. S. Chopra et al. (eds.), *Metagenomics: Techniques, Applications, Challenges and Opportunities*, https://doi.org/10.1007/978-981-15-6529-8_10

10.1 Introduction

Water is one of the most valuable resources for the existence of life, but still, a significant fraction of the population is living in water-stressed conditions. Rapid urbanization, industrialization and growing population generate huge quantity of wastewater. Many parts of the world do not have the required infrastructure to treat all the wastewater generated. Release of this untreated or partially treated wastewater is a significant cause of pathogens in receiving water bodies which leads to numerous health issues in people coming directly or indirectly in contact with it.

Sewage contaminated water contains many pathogenic viruses and microbes (Bofill-mas and Rusiñol 2020). Presence of pathogens like *Salmonella, Vibrio cholera*, bacillary, typhoid fever, *E. coli* O157:H7 in water may lead to serious, waterborne, gastrointestinal diseases (Jassim et al. 2016). Viruses present in wastewater can lead to gastroenteritis and hepatitis (Bofill-mas and Rusiñol 2020). Traditionally used bacterial indicators to assess microbial safety of water, i.e. *E. coli* and coliforms often fail to indicate pathogenic viruses which persist longer in water (Aw et al. 2014). Many human enteric viruses have been detected from environmental water samples suggesting their potentially transmission by water sources. Microbiological and virological quality of water is an essential parameter in wastewater treatment. Detection of these pathogens is essential for wastewater treatment process and public safety (Bofill-mas and Rusiñol 2020). Monitoring this transmission of viruses from contaminated environmental wastewater sources to humans is essential to determine the pathogen prevalence and its epidemiology (Haramoto et al. 2018).

Effective monitoring and control of pathogen levels is an essential component of any wastewater treatment process. Detection and characterization of pathogens in water sources is essential for its surveillance and subsequent management. It required stable techniques to detect and charcterize viruses present in wastewater. One such comparatively recent approach is to identify and characterize microbes and viruses by using metagenomics which involves sequencing and analysis of the genetic material of the sample without the need of cultivating them. Advancements in metagenomics have expanded our understanding of microbes and viruses (Hayes et al. 2017).

Many studies have demonstrated the role of metagenomics in investigating microbial and viral communities present in various environmental samples. Metagenomics has generated massive viral genome data, whose similar sequences are not present in databases. This reveals that environmental viral communities are significantly different from well-characterized viruses. This new genetic information has brought significant attention to phages as a reservoir of unexploited genetic information (Rosario and Breitbart 2011).

Numerous studies have explored bacteriophages as indicators or biocontrol agents to remove pathogenic microorganisms from wastewater. Bacteriophages also called phages are viruses (obligate intracellular parasites) that can infect and kill bacteria. Phages are the most abundant life forms in oceans and freshwaters. Previous works have explored the potential of phages as bacterial control agents and as potential biological tools in bacterial genome manipulations (Hayes et al. 2017). Their applications are recently explored in various ecosystems as in wastewater treatment as well (Wu et al. 2017).

Advanced techniques for isolation, detection and analysis of viruses will provide a better understanding of the viral nature and its interactions with the host. This will not only help us to use viruses as mere indicators of the microbial population but also as biological control agents in wastewater treatment. Discovery of novel phages that might infect pathogenic microbes may lead to the adoption of phage therapy in wastewater treatment. This chapter presents possible applications of phages in the wastewater treatment process. It also highlights the importance of metagenomic approaches in detecting and analysing viruses present in wastewater. A further role is metagenomics in other aspects of the wastewater treatment process are also discussed.

10.2 Phages: An Alternative Biological Method in Wastewater Treatment

Bactericidal properties of phage were initially employed in the medical field to control bacterial pathogens, commonly referred to as phage therapy. With time, the potential applications of the phages were explored in many areas like as food products additive, vaccines delivery vehicles, plant pathogens predators, among others. Bacteriophages can be model organisms for monitoring the microbiological quality of sludge and the effectiveness of sewage treatment plants. Bacteriophages have the potential to control the frequently encountered problems in wastewater treatment like the presence of pathogenic bacteria and the competition between the functionally important microbes and other microbes. Some studies have shown phages to be efficient in reducing the foaming in activated sludge plants (Fig. 10.1) (Withey et al. 2005).

10.2.1 Phages as Indicators

Fecal indicator bacteria and molecular techniques are routinely used to detect human viruses. Fecal coliforms are commonly used indicators of the presence of fecal contamination, but these are not reliable for prediction of enteric viruses (Ostoich et al. 2007). Removal of pathogenic human enteric viruses is an important parameter to evaluate the effectiveness of the membrane filtration. Bacteriophages can be used as alternative indicators owning to its viral nature and similarity to human enteric viruses. Recent works have proposed the utilization of indigenous bacteriophages as indicators or tracers to examine and evaluate membrane effectiveness (Wu et al. 2017).

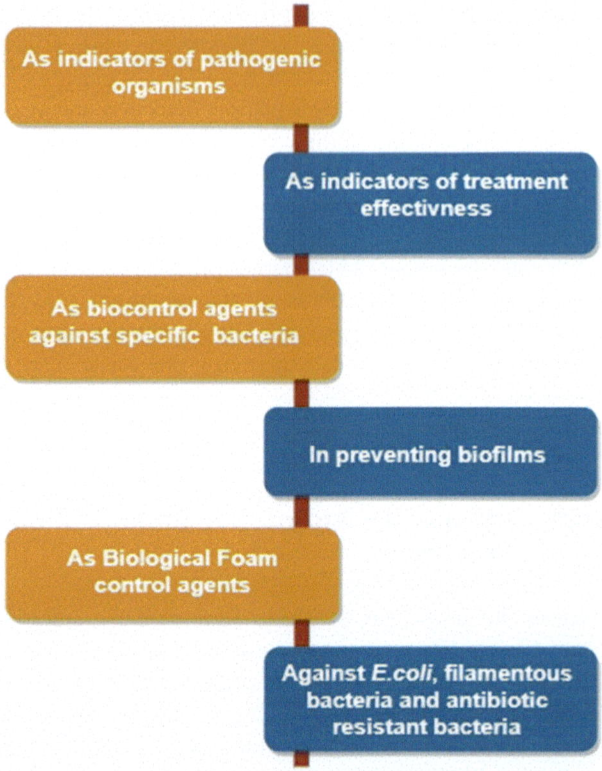

Fig. 10.1 Bacteriophages in wastewater treatment

10.2.2 Phages as Specific Biocontrol Agents

Unlike using traditional broad-spectrum antibiotics and chemicals against pathogens, phages target specific bacteria with minimal disruption to other microflora. The exploitation of phages as a biocontrol agent of bacteria has attracted much interest in various fields. Adaptability, specificity, effectiveness against their host and natural residence in the environment makes phages suitable candidates to be exploited as biocontrol agents in wastewater treatment (Jassim et al. 2016).

10.2.3 Phages to Control Filamentous Organisms in Activated Sludge Process

Activated sludge process (ASP) is one of the commonly employed techniques to treat industrial and municipal wastewater by using microorganisms. Foaming due to filamentous organisms is the main problem in this method. Formation of a thick scum on the surface of the settling sludge causes operational problems. This scum is formed by the filamentous bacterium that produces mycolic acid. Lytic phages can be used to control these filamentous bacteria thereby reducing foaming in the ASP (Jassim et al. 2016).

10.2.4 Phages in Membrane Filtration

Various membrane-based filtration processes like microfiltration, nanofiltration, ultrafiltration, forward osmosis, reverse osmosis, among others are employed in many places to treat wastewater. Some of the common issues encountered in membrane-based methods are membrane effectiveness, lifespan and fouling. Phages are also used as surrogate particles to assess membrane integrity. Owning to their antimicrobial properties, phages can be used as biological agents to control membrane fouling (Wu et al. 2017).

Successful and efficient utilization of phages in the treatment of wastewater needs a complete understanding of the microbial community dynamics and its interactions present in the wastewater (Jassim et al. 2016). Shotgun metagenomics and marker gene amplification metagenomics are primary approaches that have allowed qualitative and quantitative analysis of uncultured microbial and viral populations. Viruses are often contamination with bacterial DNA which needs to be removed prior to sequencing. Also, the lack of universal marker genes and the vast diversity in viruses have made the application of metagenomics a little tricky in virology. Nevertheless, at the same time, the small size of viruses is favorable for sequencing-based methods (Hayes et al. 2017).

10.3 Metagenomics for Detection of Viruses

Metagenomics refers to the analysis of the genetic material that has been recovered from an environmental sample to understand its diversity, structure and ecological role. It is also referred to as ecogenomics or environmental genomics. In addition to finding taxonomically and phylogenetically relevant genes, metagenomics can be used to find other novel genes (Bharagava et al. 2019). Exploration of the whole environmental genome including the uncultivable species using this approach has revolutionized the files of microbiology and virology (Bashir et al. 2014).

Conventionally culture-based methods were the standard practice for the detection of infectious viruses. With the advent of polymerase chain reaction (PCR), it quickly became a preferred method for viral identification owing to its high sensitivity and short detection time. Further advancements in PCR have allowed detection of the not easily culturable viruses, simultaneous multiple targets detection and direct multi-pathogen detection (Haramoto et al. 2018). PCR is often used in combination with plaque-forming test and molecular techniques (Hryniszyn and Skonieczna 2013).

Recovery of viruses from wastewater firstly requires concentrating viruses from the sample by using filtration based approaches like electronegative filtration, centrifugation, ultrafiltration based approaches with PEG precipitation and flocculation followed by its detection using polymerase chain reaction-based methods and infectivity assays. This has been used to reveal diverse types of viruses from different water samples like from rivers, groundwater, surface water, reclaimed water and wastewater (Bofill-mas and Rusiñol 2020). However, PCR based methods are often susceptible to inhibitory substances and do not allow differentiation of infectious from non-infectious viruses (Haramoto et al. 2018).

The availability of high-throughput sequencing and developments in broad PCRs and microarrays allowed detection of novel pathogens using sequencing data. This approach involves sequencing complete nucleic acid fragment of the sample; therefore, it is essential to remove any non-viral genome. Membrane filtration is used for removal of cells and ultrafiltration or DNase/RNase treatment for removal of free and naked nucleotides present in the sample. The viral genome is then amplified using primers with artificial sequence to generate DNA fragments for sequencing (Haramoto et al. 2018). Viral metagenomics involves shotgun sequencing of the purified viral particle (Rosario and Breitbart 2011). It has now become the standard procedure of viral detection and its genomic characterization (Fig. 10.2) (Hoeper et al. 2017).

10.4 Metagenomics Applications in Wastewater Treatment

10.4.1 Metagenomics for Pathogen Characterization

Wastewater released from industries, agriculture and other human activities pollutes soil and water. Organic and inorganic contaminants along with pathogens present in wastewater, are a severe risk to public health. Identification of pathogenic organisms is an essential component of the treatment process. Many studies have used NGS to characterize the pathogenic microbes and viruses present in the sewage treatment plants (Bharagava et al. 2019). Next-generation sequencing can be used to find out and compare the microbial and viral composition present in various wastewater treatment plants (Fig. 10.3).

Many DNA and RNA viruses have been identified till date from wastewater using viral metagenomics. It has also generated substantial novel viral genome data which do not match with nucleotide sequence databases (Haramoto et al. 2018). In addition to detection of known and unknown viruses, next-generation sequencing can be used to find and provide new data about emergent strains of the previously known viruses as well as emerging viruses in wastewater (Bofill-mas and Rusiñol 2020). These techniques also offer the opportunity of investigating the diversity of the virome to identify novel variants and emerging pathogens (Hoeper et al. 2017).

10.4.2 Metagenomics in Bioremediation

Bioremediation is another promising eco-friendly technology utilizing the ability of microbes to remediate contaminated water. It is mediated by the ability of microbes

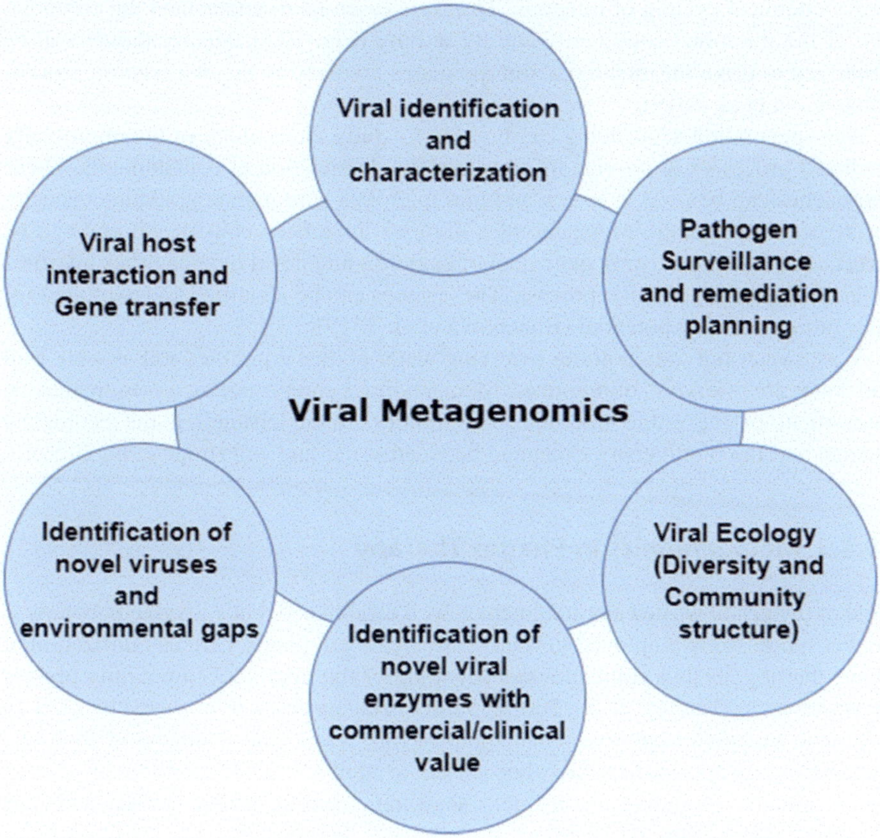

Fig. 10.2 Applications of Viral Metagenomics

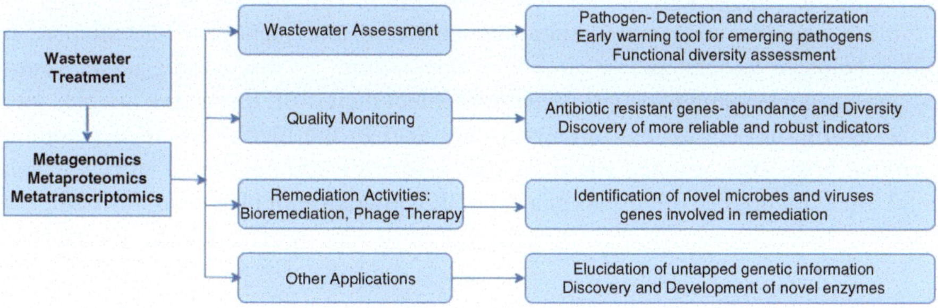

Fig. 10.3 Applications of metagenomics in wastewater treatment

to transform or degrade the contaminants present in wastewater by bioattenuation, biostimulation or bioaugmentation. Microbes are the dominant player in the transformation and degradation of pollutants. Microbes also have an essential role in the biogeochemical cycling of minerals. Thus it is essential to understand the composition of the microbial community and its activity (Fig. 10.3). Metagenomics can be employed to drive information about the genes involved in the degradation process (Bharagava et al. 2019).

Next-generation sequencing can be used to study the ecology of the microbially mediated processes to provide insights into the degradation of contaminants. These approaches can be used to screen and find microbes with high degradation capacity for a specific pollutant. Metagenomics libraries have been constructed and can be screened to find the existing gene pool of enzymes and novel gene families involved in the pollutant degradation process. These genes can be identified and exploited for their bioremediation potential (Bharagava et al. 2019).

Wastewater not only pollutes receiving water bodies it pollutes soil as well. Soil and water are reservoirs of microbes. Metagenomics can be used in a comprehensive assessment of the entire microbial community, in understanding the ecological linkage and the biodiversity exchange between water and soil (Edge et al. 2020).

10.4.3 Metagenomics in Phages Therapy

Role of phages in wastewater treatment is well established, but still, phage therapy is not yet successfully employed in wastewater treatment plants. Efficient utilization of phage therapy needs a complete understanding of the microbial community present in wastewater (Withey et al. 2005). Viral metagenomics is extensively used to catalogue viruses in environmental samples like water, soil, sludge and food etc. Metagenomics approaches have been used to study viral distribution in various water sources including freshwater, seawater, coastal water, wastewater and reclaimed water (Fig. 10.3). Many studies have demonstrated the significance of metagenomics in the characterization of viruses present in various wastewater samples (Aw et al. 2014).

Diagnostic metagenomics is increasingly finding applications in many areas by providing insights into viral community structure and pathogen surveillance. It further helps in understanding virus-host interactions by discovering novel genes and viral proteins using which viruses might infect their host. It can give insights into the co-evolution of host and virus and may lead to the identification of potential microbial hosts (Rosario and Breitbart 2011). Identification of active viruses which might infect pathogenic microbes can boost the utilization of phages in the treatment process.

10.4.4 Metagenomics to Extract Functional Information

Analysis of total mRNA, also called metatranscriptomics, can be used to extract further information about gene expression profiles of microbes present in wastewater. The recovery of protein samples known as metaproteomics can be used to find out metabolic activities and link it with the genetic diversity of microbes.

Identification and quantification of all metabolites present in a sample also called metabolomics can improve our understanding of nutrient and pollutant transformation processes (Bharagava et al. 2019).

10.4.5 Metagenomics to Find Antibiotic-Resistant Genes

Another problem area is the emergence of the wastewater treatment plants as the reservoirs of antibiotic resistance genes (ARGs) (Yin et al. 2019). The spread of antibiotic-resistant genes is a growing problem. Further emergence of resistant pathogens is a threat to human health. ARGs are usually spread through mobile genetic elements like transposons and plasmids. Various studies have attempted to detect and characterize ARGs and mobile genetic elements using NGS (Bharagava et al. 2019; Wang et al. 2013). It can reveal vital information about the abundance and diversity of ARGs present in the wastewater treatment plants (Yin et al. 2019).

10.5 Conclusion

Viruses hold diverse roles in the wastewater treatment process. Bacteriophages can be utilized as indicators of microbiological quality of wastewater. Phages can be used as specific Biocontrol agents to remove pathogenic microbes from wastewater. However, efficient utilization of this bacteriophages requires a complete understanding of microbial and viral community and its interactions in wastewater. Conventionally culture-based or molecular techniques like PCR are used for detection and characterization of wastewater. Advances in next-generation sequencing, i.e. whole genome sequencing also called metagenomics, have allowed detection of pathogens without the need to culture them. By allowing exploitation of viral and microbial communities, metagenomics has revolutionized the field of environmental virology.

Metagenomics has become an essential tool in wastewater assessment and monitoring which is a crucial component of planning the remediation activities. It has been used to monitor wastewater quality by allowing the detection and characterization of pathogenic organisms. It can be used to detect emerging pathogens as well. Metagenomics approach can be used to discover several viruses with significant and diverse functionalities. It can be used to find microbes with better pollutant degradation capacity that can be used in bioremediation. Viral metagenomics can be instrumental in the discovery of novel viral proteins or enzymes that may find applications in the remediation of wastewater. Metagenomics can reveal necessary information related to the diversity of antibiotic-resistant genes present in wastewater treatment plants as well.

Metagenomics approached related to the analysis of proteins and transcriptome can extract useful information like gene expression profiles and metabolic activities. Significant advancements in sequencing and bioinformatics have to lead to the generation of substantial viral genome data that is not yet identified. Statistical analysis and mathematical modeling are used to examine community composition and biogeography, which allows a better understanding of the viral ecology and its interaction with hosts. It is increasingly finding applications in elucidating the untapped genetic information. High-quality viral genomes can facilitate elucidation of viral evolution and also forms the basis of many future studies related to pathogen surveillance. Future work can attempt to categorize these novel viruses using third-generation sequencing technologies that can analyze thousands of nucleotides.

References

Aw TG, Howe A, Rose JB (2014) Metagenomic approaches for direct and cell culture evaluation of the virological quality of wastewater. J Virol Methods 210:15–21. https://doi.org/10.1016/j.jviromet.2014.09.017

Bashir Y, Pradeep Singh S, Kumar Konwar B (2014) Metagenomics: an application based perspective. Chinese J Biol 2014:1–7. https://doi.org/10.1155/2014/146030

Bharagava RN, Purchase D, Saxena G, Mulla SI (2019) Applications of metagenomics in microbial bioremediation of pollutants: from genomics to environmental cleanup. In: Microbial diversity in the genomic era. Elsevier, London, pp 459–477

Bofill-mas S, Rusiñol M (2020) Recent trends on methods for the concentration of viruses from water samples. Curr Opin Environ Sci Heal 16:7–13. https://doi.org/10.1016/j.coesh.2020.01.006

Edge TA, Baird DJ, Bilodeau G et al (2020) The Ecobiomics project: advancing metagenomics assessment of soil health and freshwater quality in Canada. Sci Total Environ 710:135906. https://doi.org/10.1016/j.scitotenv.2019.135906

Haramoto E, Kitajima M, Hata A et al (2018) A review on recent progress in the detection methods and prevalence of human enteric viruses in water. Water Res 135:168–186. https://doi.org/10.1016/j.watres.2018.02.004

Hayes S, Mahony J, Nauta A, Van Sinderen D (2017) Metagenomic approaches to assess bacteriophages in various environmental niches. Viruses 9:1–22. https://doi.org/10.3390/v9060127

Hoeper D, Wylezich C, Beer M (2017) Loeffler 4.0: diagnostic metagenomics. In: Advances in virus research. Elsevier, Amsterdam, pp 17–37

Hryniszyn A, Skonieczna M (2013) Methods for detection of viruses in water and wastewater. Advances in Microbiology 2013:442–449

Jassim SAA, Limoges RG, El-Cheikh H (2016) Bacteriophage biocontrol in wastewater treatment. World J Microbiol Biotechnol 32:1–10. https://doi.org/10.1007/s11274-016-2028-1

Ostoich M, Aimo E, Frate R et al (2007) Intergrated approach for microbiological impact assessment of public wastewater treatment plants. Chem Ecol 23:43–62

Rosario K, Breitbart M (2011) Exploring the viral world through metagenomics. Curr Opin Virol 1:289–297. https://doi.org/10.1016/j.coviro.2011.06.004

Wang Z, Zhang XX, Huang K et al (2013) Metagenomic profiling of antibiotic resistance genes and Mobile genetic elements in a tannery wastewater treatment plant. PLoS One 8:1–9. https://doi.org/10.1371/journal.pone.0076079

Withey S, Cartmell E, Avery LM, Stephenson T (2005) Bacteriophages--potential for application in wastewater treatment processes. Sci Total Environ 339:1–18. https://doi.org/10.1016/j.scitotenv.2004.09.021

Wu B, Wang R, Fane AG (2017) The roles of bacteriophages in membrane-based water and wastewater treatment processes: a review. Water Res 110:120–132. https://doi.org/10.1016/j.watres.2016.12.004

Yin X, Deng Y, Ma L et al (2019) Exploration of the antibiotic resistome in a wastewater treatment plant by a nine-year longitudinal metagenomic study. Environ Int 133:105270. https://doi.org/10.1016/j.envint.2019.105270

Metagenomics in Agriculture: State-of-the-Art

11

Achala Bakshi, Mazahar Moin, and M. S. Madhav

Abstract

The physical and chemical properties of soil depend upon the presence of microbial diversity. Metagenomics has potential contributions towards the study of the agriculturally important microbial population in the soil such as Plant Growth Promoting Bacteria (PGPRs). PGPRs have a huge prospective in endowing the structural and functional aspects of plant-microbial interactions followed by exploring the agronomically essential genes. The soil microbial resources can be exploited for identifying these novel genes. Metagenomics is an advanced genomic tool used to access the complete soil microbial diversity irrespective of their cultivable or non-cultivable nature. Recent molecular methods exploited for identification of desired genes are direct genomic DNA extraction, preparation of metagenomics libraries, heterologous gene expression and next-generation high throughput sequencing of environmental samples. Development of sustainable agriculture demands frequent exploitation and characterization of PGPRs. Metagenomics can considerably provide paramount of the genetic information irrespective of the culturable subsets. Furthermore, it also potentially bridges the gaps in the genetic evolution of different unidentified species of microbial communities. The Recent Next Generation Sequencing (NGS) technology enables the effective combination of crop genotyping, high throughput metagenomics and detection of the diversity of plant pathogens and PGPRs. The present chapter discusses current impact of soil microbes and advanced metagenomics technologies on plant performance and sustained crop productivity.

A. Bakshi (✉) · M. Moin · M. S. Madhav
Indian Institute of Rice Research (IIRR), Hyderabad, Telangana, India

R. S. Chopra et al. (eds.), *Metagenomics: Techniques, Applications, Challenges and Opportunities*, https://doi.org/10.1007/978-981-15-6529-8_11

Keywords

Metagenomics · Plant Growth-Promoting Rhizobacteria (PGPR) · Plant-microbe interaction · Rhizosphere · Sustainable agriculture

11.1 Introduction

The rhizosphere of plant encompasses soil-borne microbes, which contribute to the biochemical and physiological properties of the soil and influences the root growth and hence affects the plant growth and development (Hunter et al. 2014). Identification of agronomically important microbial biodiversity is still a significant task, for example, a gram of soil consists of approximately 10^{10} bacterial cells and 5×10^4 other organisms (Torsvik et al. 1990; Roesch et al. 2007; Raynaud and Nunan 2014). Also, the laboratory-culture-dependent techniques identify only 0.1–1% of microbes on growth media (Ferrari et al. 2005; Pham and Kim 2012) whereas, 99% bacteria are yet to be discovered (Goel et al. 2017). Under extreme environmental conditions, the culturable microbial population can be transformed into non-culturable species. Therefore, the culture media-dependent methods are insufficient to discover such microbes.

Furthermore, it is easy to access the genetic information of microbes existing in the environmental samples instead of growing them (Handelsman 2004; Daniel 2005). Characterization of bacterial 16S rRNA genes and fungal internal transcribed spacer 1–4, ITS1-ITS4 sequences via metagenomic pyrosequencing in the agricultural soil samples affected by drought/heatwave identified microbial assemblage linked to Carbon, Nitrogen, Phosphorus and Sulphur biogeochemical cycling under extreme environmental conditions (Acosta-Martinez et al. 2014). New molecular methods are useful in the identification of pathogenicity of microorganisms, their detection and quantification in environmental DNA samples. Mesapogu et al. (2011, 2012) identified an efficient and rapid method for detection and quantification of *Fusarium udum* fungal infection in soil samples from pigeon pea cultivation. Metagenome functions with the cloning of large environmental genomic DNA in association with high throughput sequencing methods. The sequenced metagenomic libraries contain wide information of homologous or heterologous sequences of previously identified microorganisms.

11.1.1 Role of the Microbial Community in the Plant Rhizosphere and Phyllosphere as Plant Growth Promoting Bacteria (PGPs)

Plant aerial leaf surfaces or phyllosphere harbours anoxygenic phototrophic bacterial community which displays different metabolic properties (Atamna-Ismaeel et al. 2012). Presence of phyllospheric bacteria modifies the release of leaf exudates or surfactants that can increase the water retention ability and leaf moisture by modulating stomatal closure (Hardoim et al. 2008; Mohanty et al. 2016). These bacteria produce toxins that modify leaf cuticle, resulting in the changes in ion transport of cell membranes (Schreiber et al. 2005). An endophyte association, *Rumohra adiantiformis* on leather phyllosphere causes deformation of leaves (Kloepper et al. 2013). Bacterial isolates collected from the phyllosphere of different crops such as wheat, pearl millet, cotton, mungbean and potato were characterized for various plant growth-promoting traits such as ammonia excretion, auxin (indole acetic acid) production, phosphate solubilisation and nitrate assimilation. Foliar spray of these isolates improved plant growth in terms of plant height and yield in potato (Kumari and Kumar 2018). Plant-microbe interactions in the rhizosphere are important for the development of sustainable agricultural practices and produce useful bioactive products or biofertilizers. Plant root releases exudate rich in organic carbon or nitrogen compounds in soil, These plant roots exudates help in the association of plants with microbes and also potentially provide nutrition to plants resulting in increased plant growth and disease resistance (Huang et al. 2014). Bacteria which develop their colonies in the roots of plants have the ability to stimulate plant growth and are referred to as Plant Growth Promoting Rhizobacteria (PGPR, Ahemad and Kibret 2014; Glick 2014). Different methods for energy transduction help microbes to sustain with their competitors and to make the association with allies (Gupta et al. 2018). The interaction of soil microbes with plant roots is important for nutrient recycling and carbon sink relationship (Abhilash et al. 2012). For example, rhizobacteria of Indo-Gangetic plain release tricalcium phosphates by mineral solubilisation in the soil (Bahadur et al. 2017). The soil microbial diversity in rhizosphere participates in shaping soil fertility, soil structure, plant growth, plant disease responses and also acts as bioindicators (Bramhachari et al. 2017). Plant-root associated microorganisms can influence important plant traits such as flowering time, total yield, biotic and abiotic tolerance (Van Wees et al. 2008; Marasco et al. 2012; Panke-Buisse et al. 2015). Metagenomics or use of molecular approaches is a novel method to assess the microbial population by covering a single cell to the whole microbial community in the rhizosphere (Bloem et al. 2005; Sørensen et al. 2009). Recently, the advancements in metagenomics such as whole-genome sequencing along with bioinformatic approaches not only identified novel microbes (Streit and Schmitz 2004; Hoff et al. 2008), but also the identification of genes or enzymes with PGPR ability in crops has become simple yet efficient. Functional metagenomic approaches revealed the importance of PGPR endophytic communities in rice roots (Sessitsch et al. 2012) and the presence of ammonia-oxidizing archaea or bacteria in soil (Schauss et al. 2009). Some bacterial genera including *Rhizobium, Thiobacillus, Azospirillum, Agrobacterium, Burkholderia, Frankia, Pseudomonas* and *Bacillus,* are considered as PGPRs with plant-growth-promoting and biocontrol activities (Vessey 2003; Ramesh et al. 2009). Rhizospheric bacteria also harbours adaptive traits to plants under abiotic and biotic stresses and can potentially increase plant immunity by inducing Systemic Resistance (Yang et al. 2009). PGPR increases osmolyte accumulation and phytohormone signalling under abiotic stress. *Arabidopsis thaliana* transgenic plants constitutively expressing *Bacillus subtilis* proBA gene enhances proline synthesis to confer salt tolerance (Chen et al. 2007). Inoculation of SN23 and

SQR9 isolates of *Bacillus amyloliquefaciens* in rice and maize respectively confers tolerance to salinity stress in both hydroponic and normal conditions and simultaneously upregulated the genes related to osmotic and ionic stress (Nautiyal et al. 2013; Chen et al. 2016). Tillage systems in agriculture land use also change the structure and composition of soil microbiome (Carbonetto et al. 2014). Priming rhizosphere bacteria in wheat seedlings induced drought stress tolerance with an increase in plant biomass (Timmusk et al. 2014). Association of AR156 from *Bacillus cereus*, SM21 from *Bacillus subtilis*, and XY21 from *Serratia* sp. were found to have biocontrol and drought-tolerant properties in cucumber plants (Wang et al. 2012). *Trichoderma harzianum* elicits antioxidant defense mechanisms in tomato seedlings to provide tolerance to water deficit conditions by enhancing reactive oxygen species (ROS)-scavenging and oxidation of ascorbate and glutathione (Mastouri et al. 2012). PGPR also ameliorated the salinity (NaCl) stress in tomato plants and solubilized the phosphates (Tank and Saraf 2010). Soil bacteria can solubilize phosphorus (P) and potassium (K) for increasing crop yields (Nath et al. 2017). Rhizobacteria can effectively produce siderophore, an iron-chelating compound of low molecular weight and increase the solubility of ferric ions in the soil (Pahari and Mishra 2017). Furthermore, the metagenomic analysis of rhizospheric microbial communities of grass family crops such as rice, maize, sorghum, among others, is necessary for providing a method to attain the important yield traits and stress tolerance to plants without any genetic manipulation. Also, the identification of the microbial community with metal sequestration or resistance genes or contaminant degrading genes or enzymes would be useful for transforming and restoring the contaminated lands with reduced toxicity.

11.2 Environmental Sampling and Extraction of Metagenomic DNA

Metagenomic DNA can be extracted by two methods: (1) direct and (2) indirect method. In direct extraction, the DNA is separated from the cell debris by direct cell wall lysis in the environment sample (Ogram et al. 1987). This technique results in a mixture of lesser quality sheered DNA of prokaryotic and eukaryotic cells (Gabor et al. 2003). A high amount of humic acid content in soil samples is a major constraint for soil DNA isolation that inhibits enzymatic reactions (Latha et al. 2009). In the indirect method, the cells are separated first from the samples and then lysed by enzymes, chemicals or mechanical disruption (Holben et al. 1988). Mechanical disruption of the cells includes thermal shocks or freeze-thaw shocks, ultrasonication and bead homogenization (Gupta et al. 2018).

11.2.1 Construction and Screening of a High-Quality Metagenomic Library

The metagenomes are fragmented for preparation of genomic libraries of different sizes and then cultured into a suitable host. Metagenomic libraries are prepared by cloning high molecular weight DNA into different plasmid vectors such as for high molecular weight DNA fragments bacterial artificial chromosome (BAC) or fosmid or cosmid vectors are used, and the cloned genes are expressed into a suitable host such as *Escherichia coli*, *Agrobacterium tumefaciens*, and *Saccharomyces cerevisiae* (Rondon et al. 2000; Craig et al. 2010; Delmont et al. 2011; Ekkers et al. 2012). Production of some secondary metabolites causes phenotypic changes in host cells like pigmentation, which can be easily detected and screened by colorimetric methods. Another change is antibiosis, which is a property of inhibiting the growth of other organisms (Craig et al. 2010). Quenching and quorum sensing methods by cloning metagenome into a promoter trap vectors fused with reporter fluorescence protein are used for antibiotic development (Hao et al. 2010; Lee et al. 2011; Romero et al. 2011). The purified and confirmed libraries are then sequenced using high throughput sequencing methods.

11.2.2 Sequencing and Analysis of Metagenomic Libraries

Sanger sequencing, in combination with 16SrRNA or ITS phylogenetic analysis, was used to identify clones with heterologous gene expression (Klindworth et al. 2013). Identification of random flanked sequences adjacent to the known conserved sequences is a substitute for the cloning-based methods, which require a suitable host for cloned gene expression (Mesapogu et al. 2012). The molecular-based techniques such as DGGE (Denaturing Gradient Gel Electrophoresis), TGGE (Temperature Gradient Gel Electrophoresis), TTGGE (Temporal Temperature Gradient Gel Electrophoresis), T-RFLP (Terminal-Restriction Fragment Length Polymorphism), SSCP (Single-Stranded Conformation Polymorphism), RAPD (Random Amplified Polymorphic DNA), ARDRA (Amplified Ribosomal DNA Restriction Analysis), real-time qRT-PCR (Fischer and Lerman 1983; Fritsch and Rieseberg 1996; Konstantinidis and Tiedje (2007); Smith and Osborn 2009) and secondary metabolite based techniques such as PLFA (PhosphoLipid Fatty Acids), CLPP (Community Level Physiological Profile) analysis using BIOLOG Ecoplates (Biolog Inc., USA) are culture-independent techniques for identification of microbial communities in the environmental samples (Fig. 11.1; Grayston et al. 2004; Nocker et al. 2007). These methods provide less information and are costly, whereas (Next Generation Sequencing) NGS methods such as ABI SOLiD, Helioscope, Solexa, 454 FLX does not require bacterial cloning. In contrast, the cloning-dependent methods such as substrate-induced gene expression (SIGEX) and the product-induced gene expression (PIGEX) approaches are cost-efficient and produce sufficient data as an outcome (Kircher and Kelso 2010; Oulas et al. 2015). Next Generation whole-genome sequencing is the best method for identifying novel molecular markers

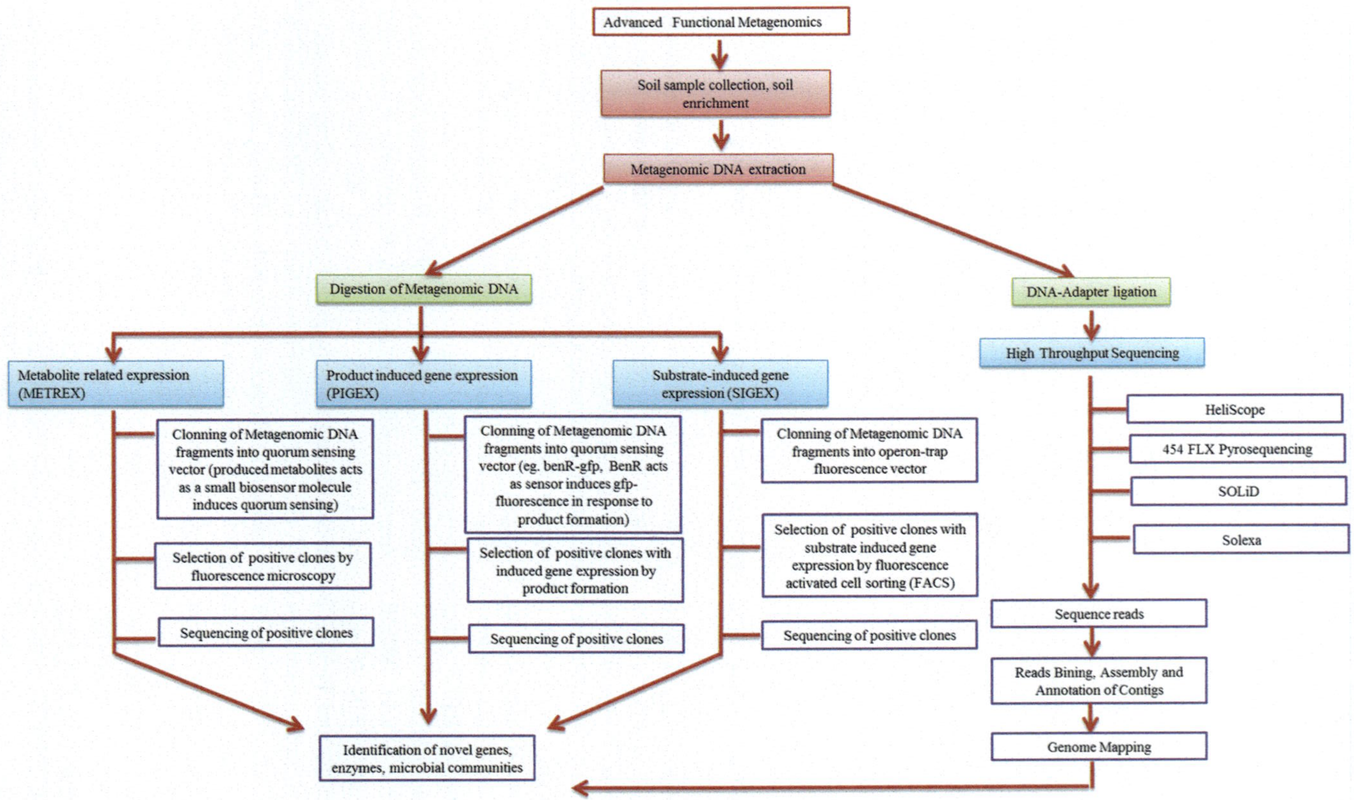

Fig. 11.1 Representation of modern advanced techniques in metagenomics used to identify novel genes in agriculture

and in studies associated with the genetic mapping in crops. Metabolite related expression (METREX) is a method that uses quorum sensing pathway for the expression of green fluorescent protein (GFP; Tyson et al. 2004; Williamson et al. 2005; Fig. 11.1).

Sequenced data are then assembled and analysed using several bioinformatics tools. Metagenomic sequences are analysed for identifying gene of interest or putative genes with functions of interest and are annotated according to their taxonomic properties. Softwares, Newbler (Roche), AMOS or MIRA, are implicated for reference-based assembly of sequences (Miller et al. 2010). Meta Velvet and Meta-IDBA software are de-novo assemblers of metagenomic sequencing data (Peng et al. 2011). DNA sequences are sorted and grouped based on their taxonomic groups called binning of Metadata (Diaz et al. 2009). S-GSOM, IMG/M, MEGAN, MG-RAST, MOTHUR, TANGO, CARMA, PCAHIER, Phylopythia, SOrt-ITEMS, TACAO, MetaCluster, MetaPhyler, and PhymmBL are algorithms used for binning metagenome sequenced data (Huson et al. 2007; Chan et al. 2008; Brady and Salzberg 2009; Diaz et al. 2009; Haque et al. 2009; Glass et al. 2010; Markowitz et al. 2012).

Next-generation sequencing of metagenomic samples identifies functional and taxonomical differences between microbial communities. The sequenced metagenomic datasets are carefully analysed statistically. Several statistical techniques such as Multidimensional Scaling (MDS), Random Forest (RF) and Linear Discriminant Analysis (LDA) are implicated for functional analysis and are publicly available metagenomic datasets (Akond et al. 2016). The first dimension MDS analysis separates the human/animal microbes from other samples and the second dimension separates them from the aquatic and mat communities (Akond et al. 2016). Similarly, the LDA plots are used to compare the samples with the human and animal-associated samples, the other microbial mats, along with the aquatic samples. RF analysis is useful for identifying differences between free-living microbes and host-associated microbes (Akond et al. 2016). Primer-E-Package performs a range of multivariate analysis. In contrast, some web-based tools such as analysis of similarities (ANOSIM), species, and identification of gene functions (SIMPER) are used for multidimensional scaling (Hughes et al. 2001; White et al. 2009). Shotgun-functionalizer is used for evaluating functional differences between data. A large number of datasets in the metagenomic analysis required informative databases for comparative functional analysis.

11.3 Identification of Agronomically Important Genes and Pathways by Metagenomics

Metagenomics is an essential tool in biotechnology which is being used in the discovery of novel genes, enzymes and bioactive molecules involved in biochemical pathways (Alves et al. 2018). The conserved 16SrRNA gene in the metagenome is used as a taxonomic marker (Fig. 11.2, Table 11.1).

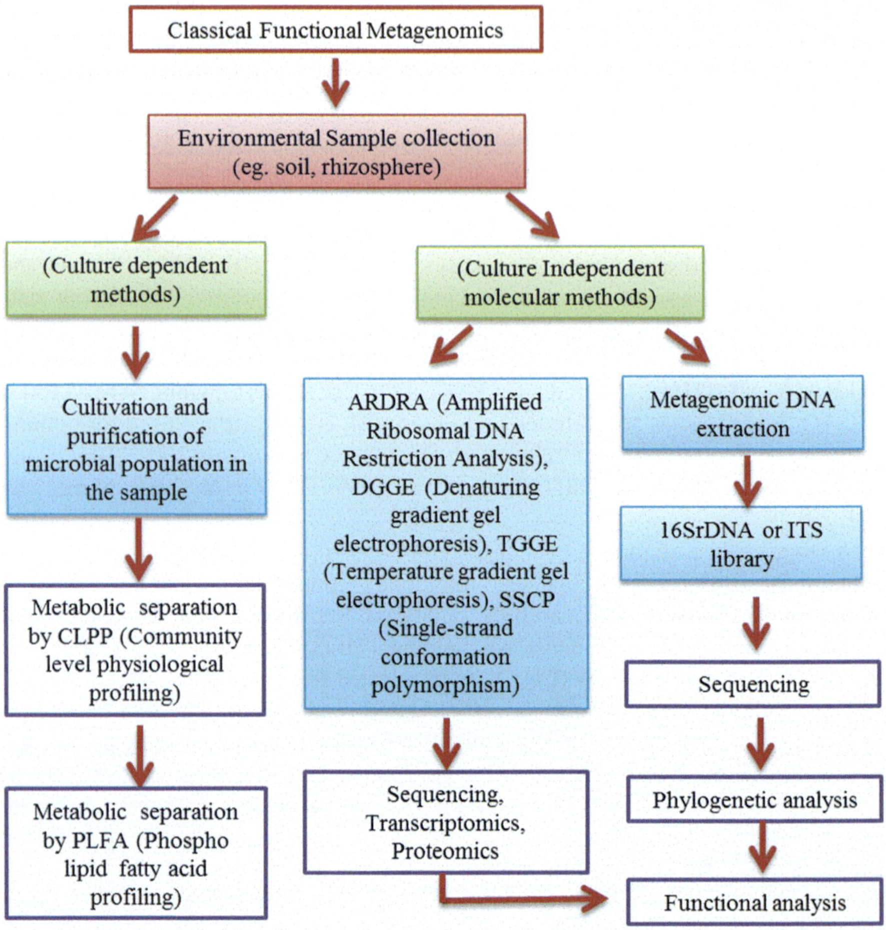

Fig. 11.2 Representation of Classical functional metagenomics with culture-dependent and culture-independent methods

Phytic acid is an organic form of phosphate in the soil which remains unutilized by the plants. Soil microbes in the rhizosphere contribute towards the phytic-acid-utilization by converting it into plant-available inorganic phosphorus (Unno and Shinano 2012). Agriculture soil from the wheat cultivated farm has identified a novel histidine acid-phosphatase family, phytase (Tan et al. 2014). Several groups have identified the mineral phosphate solubilizing traits from barley rhizosphere (Chhabra et al. 2013). PGPR increases nutrient availability in soil and also produces biologically active compounds such as phytohormones (auxins and Indole acetic acid (IAA)) for promoting plant growth. Certain PGPR contains ACC-deaminase (aminocyclopropane-1-carboxylate-deaminase) enzyme, which hydrolyzes ACC into ammonia and alpha-ketobutyrate. Similar genes encoding for ACC-deaminase were isolated and sequenced from soil rhizobacteria (Govindasamy et al. 2008;

Table 11.1 Identification of novel genes, enzymes, biocatalysts and other agriculturally important products by functional metagenomics in the past 20 years

S. no.	Nature of sample	Gene name and function	References
1.	Wood-feeding higher termite	Polysaccharide-degrading gene	Liu et al. (2019)
2.	Acidic peatland	Genes for sulfur metabolism	Hausmann et al. (2018)
3.	Soil of a northern Californian grassland	Genes related to secondary metabolite biosynthesis	Crits-Christoph et al. (2018)
4.	Anoxic lake sediments	Toluene-producing enzyme PhdB	Beller et al. (2018)
5.	Arctic glacier forefields	Genes for nitrogenous fixation	Nash et al. (2018)
6.	Marine and terrestrial environments	Genes for sulfate/sulfite reduction	Anantharaman et al. (2018)
7.	Phosphorus-deficient and phosphorus-rich soil	Genes or enzymes related to carbon, nitrogen, phosphorus and sulfur cycle	Yao et al. (2018)
8.	Rhizobacteria at Fly ash dumps	Indole acetic acid	Malhotra et al. (2017)
9.	High-latitude cold coastal environment samples	Hydrocarbon degradation genes	Espínola et al. (2018)
10.	Soil samples	Halo- and thermotolerant Enzyme, Cellulase	Garg et al. (2016)
11.	Soil samples	Genes conferring tolerance to lignocellulose-derived inhibitors	Forsberg et al. (2016)
12.	Brines and moderate salinity rhizosphere	Salt resistance genes	Mirete et al. (2015)
13.	Acid mine drainage	Arsenic resistance genes	Morgante et al. (2015)
14.	Marine water	Carboxylesterase, cold-active and salt-resistant enzyme	Tchigvintsev et al. (2015)
15.	Artificially polluted soil/water and contaminated oil	Oxygenases	Nagayama et al. (2015)
16.	Agriculture soil	Histidine acid phosphatase family, novel phytase genes	Tan et al. (2014)
17.	Spring water	Thermotolerant, heath-active andhydrothermal enzyme, β-glucosidase	Schröder et al. (2014)
18.	Plankton and rhizosphere	Genes involved in acid resistance eg. *ClpXP* protease	Guazzaroni et al. (2013)
19.	Wastewater treatment plant	Phenol hydroxylases and catechol 2,3-dioxygenases, used for degradation of aromatic compound	Silva et al. (2013)
20.	Barley rhizosphere soil	Mineral phosphate solubilizing genes	Chhabra et al. (2013)

(continued)

Table 11.1 (continued)

S. no.	Nature of sample	Gene name and function	References
21.	Rhizosphere	Phytic-acid-utilizing genes	Unno and Shinano (2012)
22.	Leaf-branch compost	Cutinases, potential use in polyethylene terephthalate (PET) degradation	Sulaiman et al. (2012)
23.	Copper-polluted agricultural soils	Copper-resistant genes	Altimira et al. (2012)
24.	Soil pest, Serratia entomophila	Toxicity for Phyllophaga blanchardi larvae	Rodríguez-Segura et al. (2012)
25.	Genome sequencing of Penicillium coprobium PF1169	Cytochrome P450 monooxygenase genes, pyripyropene A (PyA), inhibitors of acyl-CoA:Cholesterol acyltransferase	Hu et al. (2011)
26.	Wheat rhizosphere	*ACC deaminase* gene	Govindasamy et al. (2008)
27.	Rhizosphere	Genes for nickel resistance	Mirete et al. (2007)
28.	Oil-contaminated soil	Naphthalene dioxygenase	Ono et al. (2007)
29.	Activated sludge	Extradiodioxygenases	Suenaga et al. (2007)
30.	*Rhodococcus opacus* TKN14 isolated from soil contaminated with o-xylene	o-xylene oxygenase	Maruyama et al. (2005)
31.	*Phytophthora nicotianae*	HSP genes	Shan and Hardham (2004)
32.	*Ustilago hordei*	*UhAvr1*, avirulence gene	Linning et al. (2004)
33.	Leaves of strawberry cultivar, *Alternaria alternata*	AF toxin genes, host-specific toxin synthesis genes	Ito et al. (2004)
34.	*Tribolium castaneum* larva	*TcCHS1* and *TcCHS2*, chitin synthase	Arakane et al. (2004)
35.	Water lakes	Cellulase/esterase	Rees et al. (2003)
36.	*Pseudomonas syringae* inoculated on tobacco leaves	*ShcA*, a secretion genes	van Dijk et al. (2002)
37.	Phytophthora infestans inoculated on potato plant	Genes conferring avirulence to oomycete plant pathogen	Whisson et al. (2001)
38.	Xanthomonas *oryzae* pv. *pryzae*, inoculated on rice plants	Virulence genes, avirulence, hypersensitivity, pathogencity genes	Ochiai et al. (2001)

(continued)

Table 11.1 (continued)

S. no.	Nature of sample	Gene name and function	References
39.	*Xenorhabdus nematophilus* PMF1296 obtained from NCIMB culture collection and isolated from soil	Insecticidal protein against Pieris brassicae larvae	Morgan et al. (2001)
40.	*Xanthomonas oryzae* pv. *oryzae*	Virulence and xylanase activity genes	Ray et al. (2000)

Ilangumaran and Smith 2017). Molecular analysis of samples collected from extreme environment conditions identified important genes such as thermostable heat-active archaeal β-glucosidase from a hydrothermal spring metagenome (Schröder et al. 2014) and novel acid resistance genes from the extremely acidic metagenome of the Tinto River (Guazzaroni et al. 2013).

11.3.1 Metagenomics as a Tool for Weed Management

Weeds and crops compete for soil nutrition, resulting in major yield loss in crops. Therefore, prevention of weeds is necessary for crop cultivation in sustainable agriculture. Agriculture weeds are mostly adapted to extreme environments and can thrive well under changing environmental conditions (Smith 2015). In agricultural practices, crop rotation and applications of herbicides are used for weed management where herbicides can also affect crop growth and compromises the total crop productivity. To avoid this, herbicide-resistance genetically modified crops have been generated that induces resistance to herbicides, glyphosate and 2, 4-D. Herbicidal compounds secreted by soil microbes suppress germination of weed seedlings by inhibiting early growth of roots (Weissmann et al. 2003; Medd and Campbell 2005; Zdor et al. 2005; Einhorn and Brandau 2006; Banowetz et al. 2008). Actinomycetes such as Streptomyces species and other bacterial species such as pseudomonads are potent producers of herbicidal compounds (Barazani and Friedman 2001; Culliney 2005). Application of *Pseudomonas fluorescens* inhibits germination of goatgrass, *Aegilops cylindrica* (Kennedy and Stubbs 2007; Armstrong et al. 2009) and *Streptomyces* species produce methoxy-hygromycin antibiotic with monocot weed suppression activity (Lee et al. 2003). Other commercial herbicides such as glufosinate or phosphinothricin, commercially known as Basta®, earlier known as Bialaphos was first isolated from *Streptomyces viridochromogenes* and *Streptomyces hygroscopicus* (Hoerlein 1994). Basta® is a nonselective herbicide that suppresses a wide range of monocot and dicot plants by inhibiting glutamine synthetase. Similarly, thaxtomin A and tagetitoxin herbicides are isolated from *Streptomyces acidiscabies* and *Pseudomonas syringae* pv. *tagetis*, respectively, which are highly selective for monocot and dicot weeds (Strange 2007; Lydon et al. 2011). Recent understandings on plant-associated soil microbes have led to the establishment of new methods in weed prevention and management

(Trognitz et al. 2016). The current status of growing weed population and its increasing resistance to global herbicides indicates the need to develop new, improved herbicides (Heap 1999; Kao-Kniffin et al. 2013). The metagenomic based identification of new herbicides or weed-suppressive compounds is necessary that can significantly contribute to sustainable agriculture. Metagenomic functional approaches are helpful in the identification of novel herbicides by cloning metagenomic DNA into a vector expressing genes for intermediate compounds of biosynthetic pathways. Similarly, herbicide resistance genes can also be identified by screening metagenomic clones on high herbicide concentration (Kao-Kniffin et al. 2013). Function-based screening of the expression host combined with the sequence-based techniques led to the discovery of many novel natural compounds with herbicidal or pesticidal activity.

11.3.2 Metagenomics as a Tool for Pest Management

Microbial derived compounds are useful for pest management and as a biocontrol activity in agricultural systems. *Pieris brassicae*, white butterfly causes a huge loss in the productivity of cabbage crop. Molecular sequencing of *Xenorhabdus nematophilus* genome identified genes encoding for proteins related to insecticidal activity, these have been isolated and used as a pathogen for *P. brassicae* (Morgan et al. 2001). Similarly, a bacterial strain, Mor4.1 of *Serratia entomophila* have been identified as a biocontrol for several soil pests (Rodríguez-Segura et al. 2012). Metagenomic approaches have been used for the development of plant pathogenicity related diagnostics and their control. Some examples are the identification of virulence genes in *Xanthomonas oryzae* pv. *oryzae*, *Phytophthora infestans*, *Ustilago hordei*, *Xanthomonas campestris* pv. *vesicatoria* (Ochiai et al. 2001; Ray et al. 2000; Whisson et al. 2001; Linning et al. 2004).

11.4 Future Prospects of Metagenomic Research in Agriculture

The properties of plant-associated microbes provide the nutrients to the plants that are required for their suitable growth, thus helps in sustainable agricultural productivity. Bioprospecting of agronomically important bioactive compounds such as secondary metabolites, biocontrol products and stress-responsive chemicals are relevant for crop protection and improvement (Müller et al. 2016). Comprehensive screening of microbial populations or pure strains resulted in the identification of weed-suppressing compounds (Kao-Kniffin et al. 2013). Moreover, the development of rapid and cheap methods for obtaining vast genetic information is need of metagenomic studies. Application of metagenomics in the discovery of natural compounds in sustainable agriculture is based on vector-host expression system. To avoid this limitation, more advanced methods are needed to be developed for screening a large number of clones. Metagenomics potentially deciphers genetic information of microbial populations inhabiting in extreme environments such as low and high temperatures, saline and acidic conditions (Guazzaroni et al. 2013;

Schröder et al. 2014). The role of these genes that provided tolerance to extreme environmental stresses may be exploited in plant systems and be used in crop improvement programs in agriculture. Despite all benefits and limitations, undoubtedly metagenomics has provided information on novel microbial genes and pathways that allows them to thrive in different environments. Molecular metagenomics, along with the combination of sequence-based approaches, can help in understanding the properties of microbial communities and in unravelling their underlying processes or pathways. Although a vast number of studies are conducted in the past few decades, it is still necessary to develop novel bioinformatics tools for the analysis of huge metagenomic data. Also, the function-based screening of metagenomic clones is an efficient method but is based on heterologous gene expression in the host-vector system. Therefore, it is necessary to develop more specific vectors for identifying novel genes and biosynthetic pathways. Processing and storage of metagenomic data is another criterion among researchers and to overcome this concern. A storage system is further needed to develop for large datasets. Identification of single genome with the assimilation of a labelled compound can confer the identification of non-cultivated microbes of agronomical importance. DNA-SIP (DNA stable isotope probing) method is used for the detection of active microbes in the environment. These microbes are identified based on their ability to uptake or incorporate different isotopically labelled substrates (Malmstrom and Eloe-Fadrosh 2019). Metagenomics has a prominent role in the development of an advanced CRISPR/Cas9 technology in recent years (Barrangou et al. 2007 & Barrangou and Van Der Oost 2013). This technology enables the targeted genome modification in a wide range of eukaryotes and is developed as a new genome-editing tool in agriculture. CRISPR (clustered regularly interspaced short palindromic repeats) associated RNA-guided endonucleases known as Cas9 can be targeted to the specific genomic location with the help of a short RNA guide (Barrangou et al. 2007). A new CasX enzyme is identified recently in Delta-proteobacteria (DpbCasX) from groundwater samples using metagenomic approaches (Song 2019). The unique non-target strand-binding (NTSB) and target-strand-loading (TSL) domain of DpbCasX help in bending an RNA–DNA hybrid to facilitate the cleavage of a targeted DNA strand (Song 2019). The CRISPR/Cas technology has potentially enhanced the breeding methods in sustainable agriculture with the desired beneficial traits in the elite crop varieties.

11.5 Conclusion

Enhanced crop productivity is a basic requirement to fulfil the increasing demands of staple food crops in developing countries. To gain huge production in sustainable agriculture, genetic improvement is needed in crops with increased resistance to climate changes and soil nutrition quality or soil fertility. The microbial communities present in the surrounding environment of a plant such as a phyllosphere or rhizosphere, act as natural resource in increasing crop health and providing resistance to various diseases and environmental factors. Furthermore, NGS technology

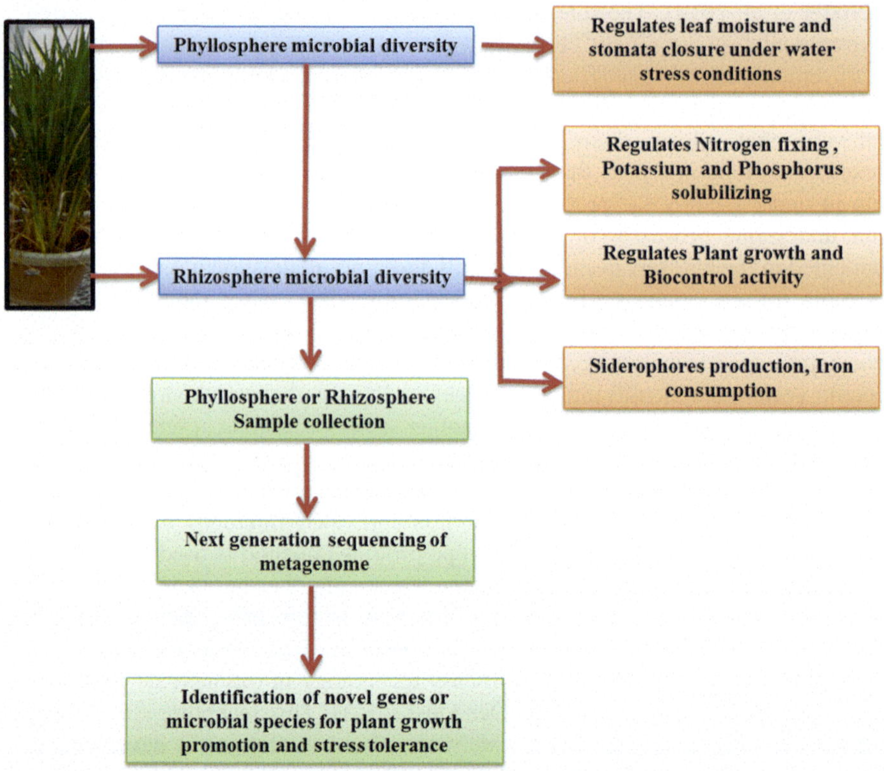

Fig. 11.3 Illustration of the importance of metagenomics in sustainable agriculture and crop improvement

can be extended to the discovery and characterization of novel agronomically important microbes (Fig. 11.3).

The combination of high throughput next-generation technologies along with metagenomic approaches, can harness the knowledge of unknown plant–microbiome interactions to identify agronomically beneficial microbial communities present in the plant rhizosphere. Furthermore, the knowledge of unrevealed plant-microbial interactions can develop a future improved plant breeding in the agricultural system.

Acknowledgements AB acknowledges financial support from the Department of Biotechnology-Research Associate program in Biotechnology and Life Sciences (2-29/RA/Bio/2018/550), and Department of Biotechnology, ICAR-Indian Institute of Rice Research, Hyderabad.

References

Abhilash PC, Powell JR, Singh HB, Singh BK (2012) Plant–microbe interactions: novel applications for exploitation in multipurpose remediation technologies. Trends Biotechnol 30:416–420

Acosta-Martinez V, Cotton J, Gardner T et al (2014) Predominant bacterial and fungal assemblages in agricultural soils during a record drought/heat wave and linkages to enzyme activities of biogeochemical cycling. Appl Soil Ecol 84:69–82

Ahemad M, Kibret M (2014) Mechanisms and applications of plant growth promoting rhizobacteria: current perspective. J King saud Univ 26:1–20

Akond Z, Alam M, Ahmed MS, Mollah MNH (2016) Multivariate statistical techniques for metagenomic analysis of microbial community recovered from environmental samples. J Bio-Sci. 24:45–53

Altimira F, Yáñez C, Bravo G et al (2012) Characterization of copper-resistant bacteria and bacterial communities from copper-polluted agricultural soils of Central Chile. BMC Microbiol 12:193

Alves LF, Westmann CA, Lovate GL, de Siqueira GMV, Borelli TC, Guazzaroni ME (2018) Metagenomic Approaches for Understanding New Concepts in Microbial Science. Int J Genomics. 2018:2312987

Anantharaman K, Hausmann B, Jungbluth SP et al (2018) Expanded diversity of microbial groups that shape the dissimilatory sulfur cycle. ISME J 12:1715–1728

Arakane Y, Hogenkamp DG, Zhu YC et al (2004) Characterization of two chitin synthase genes of the red flour beetle, *Tribolium castaneum*, and alternate exon usage in one of the genes during development. Insect Biochem Mol Biol 34:291–304

Armstrong D, Azevedo M, Mills D et al (2009) Germination-arrest factor (GAF): 3. Determination that the herbicidal activity of GAF is associated with a ninhydrin-reactive compound and counteracted by selected amino acids. Biol Control 51:181–190

Atamna-Ismaeel N, Finkel O, Glaser F et al (2012) Bacterial anoxygenic photosynthesis on plant leaf surfaces. Environ Microbiol Rep 4:209–216

Bahadur I, Maurya BR, Meena VS et al (2017) Mineral release dynamics of tricalcium phosphate and waste muscovite by mineral-solubilizing rhizobacteria isolated from indo-gangetic plain of India. Geomicrobiol J 34:454–466

Banowetz GM, Azevedo MD, Armstrong DJ et al (2008) Germination-arrest factor (GAF): biological properties of a novel, naturally-occurring herbicide produced by selected isolates of rhizosphere bacteria. Biol Control 46:380–390

Barazani O, Friedman J (2001) Allelopathic bacteria and their impact on higher plants. Crit Rev Microbiol 27:41–55

Barrangou R, Fremaux C, Deveau H et al (2007) CRISPR provides acquired resistance against viruses in prokaryotes. Science 80(315):1709–1712

Barrangou R, Van Der Oost J (2013) CRISPR-Cas systems. RNA-mediated adapt Immun Bact archaea, 101007978364234657 6th edn. Heidelberg SVB, Ed

Beller HR, Rodrigues AV, Zargar K et al (2018) Discovery of enzymes for toluene synthesis from anoxic microbial communities. Nat Chem Biol 14:451–457

Bloem J, Hopkins DW, Benedetti A (2005) Microbiological methods for assessing soil quality. CABI, Cambridge

Brady A, Salzberg SL (2009) Phymm and PhymmBL: metagenomic phylogenetic classification with interpolated Markov models. Nat Methods 6:673–676

Bramhachari PV, Nagaraju GP, Kariali E (2017) Metagenomic approaches in understanding the mechanism and function of PGPRs: perspectives for sustainable agriculture. In: Agriculturally Important Microbes for Sustainable Agriculture. Springer, Singapore, pp 163–182

Carbonetto B, Rascovan N, Alvarez R et al (2014) Structure, composition and metagenomic profile of soil microbiomes associated to agricultural land use and tillage systems in Argentine Pampas. PLoS One 9:e99949

Chan C-KK, Hsu AL, Halgamuge SK, Tang S-L (2008) Binning sequences using very sparse labels within a metagenome. BMC Bioinformatics 9:215

Chen L, Liu Y, Wu G, Njeri VK, Shen Q, Zhang N, et al (2016) Induced maize salt tolerance by rhizosphere inoculation of Bacillus amyloliquefaciens SQR9. Physiol Plant 158:34–44

Chen M, Wei H, Cao J et al (2007) Expression of *Bacillus subtilis* proBA genes and reduction of feedback inhibition of proline synthesis increases proline production and confers osmotolerance in transgenic *Arabidopsis*. BMB Rep 40:396–403

Chhabra S, Brazil D, Morrissey J et al (2013) Characterization of mineral phosphate solubilization traits from a barley rhizosphere soil functional metagenome. Microbiology 2:717–724

Craig JW, Chang F-Y, Kim JH et al (2010) Expanding small-molecule functional metagenomics through parallel screening of broad-host-range cosmid environmental DNA libraries in diverse proteobacteria. Appl Environ Microbiol 76:1633–1641

Crits-Christoph A, Diamond S, Butterfield CN et al (2018) Novel soil bacteria possess diverse genes for secondary metabolite biosynthesis. Nature 558:440–444

Culliney TW (2005) Benefits of classical biological control for managing invasive plants. CRC Crit Rev Plant Sci 24:131–150

Daniel R (2005) The metagenomics of soil. Nat Rev Microbiol 3:470–478

Delmont TO, Robe P, Cecillon S et al (2011) Accessing the soil metagenome for studies of microbial diversity. Appl Environ Microbiol 77:1315–1324

Diaz NN, Krause L, Goesmann A et al (2009) TACOA–taxonomic classification of environmental genomic fragments using a kernelized nearest neighbor approach. BMC Bioinformatics 10:56

van Dijk K, Tam VC, Records AR et al (2002) The ShcA protein is a molecular chaperone that assists in the secretion of the HopPsyA effector from the type III (Hrp) protein secretion system of *Pseudomonas syringae*. Mol Microbiol 44:1469–1481

Einhorn G, Brandau J (2006) Influence of microorganisms on weed and crop seeds? J PLANT Dis Prot 20:317–324

Ekkers DM, Cretoiu MS, Kielak AM, van Elsas JD (2012) The great screen anomaly—a new frontier in product discovery through functional metagenomics. Appl Microbiol Biotechnol 93:1005–1020

Espínola F, Dionisi HM, Borglin S et al (2018) Metagenomic analysis of subtidal sediments from polar and subpolar coastal environments highlights the relevance of anaerobic hydrocarbon degradation processes. Microb Ecol 75:123–139

Ferrari BC, Binnerup SJ, Gillings M (2005) Microcolony cultivation on a soil substrate membrane system selects for previously uncultured soil bacteria. Appl Environ Microbiol 71:8714–8720

Fischer SG, Lerman LS (1983) DNA fragments differing by single base-pair substitutions are separated in denaturing gradient gels: correspondence with melting theory. Proc Natl Acad Sci 80:1579–1583

Forsberg KJ, Patel S, Witt E et al (2016) Identification of genes conferring tolerance to lignocellulose-derived inhibitors by functional selections in soil metagenomes. Appl Environ Microbiol 82:528–537

Fritsch P, Rieseberg, LH (1996) The use of random amplified polymorphic DNA (RAPD). In: Smith TB, Wayne, RK (eds) Conservation genetics, molecular genetic approaches inconservation. London, Oxford Univ. Press, pp 54–73

Gabor EM, de Vries EJ, Janssen DB (2003) Efficient recovery of environmental DNA for expression cloning by indirect extraction methods. FEMS Microbiol Ecol 44:153–163

Garg R, Srivastava R, Brahma V et al (2016) Biochemical and structural characterization of a novel halotolerant cellulase from soil metagenome. Sci Rep 6:39634

Glass EM, Wilkening J, Wilke A et al (2010) Using the metagenomics RAST server (MG-RAST) for analyzing shotgun metagenomes. Cold Spring Harb Protoc 2010:pdb-prot5368

Glick BR (2014) Bacteria with ACC deaminase can promote plant growth and help to feed the world. Microbiol Res 169:30–39

Goel R, Suyal DC, Dash B, Soni R (2017) Soil metagenomics: a tool for sustainable agriculture. In: Mining of microbial wealth and metagenomics. Springer, Singapore, pp 217–225

Govindasamy V, Senthilkumar M, Gaikwad K, Annapurna K (2008) Isolation and characterization of ACC deaminase gene from two plant growth-promoting rhizobacteria. Curr Microbiol 57:312–317

Grayston SJ, Campbell CD, Bardgett RD et al (2004) Assessing shifts in microbial community structure across a range of grasslands of differing management intensity using CLPP, PLFA and community DNA techniques. Appl Soil Ecol 25:63–84

Guazzaroni M, Morgante V, Mirete S, González-Pastor JE (2013) Novel acid resistance genes from the metagenome of the Tinto River, an extremely acidic environment. Environ Microbiol 15:1088–1102

Gupta N, Vats S, Bhargava P (2018) Sustainable agriculture: role of metagenomics and metabolomics in exploring the soil microbiota. In: In Silico Approach for Sustainable Agriculture. Springer, Singapore, pp 183–199

Handelsman J (2004) Metagenomics: application of genomics to uncultured microorganisms. Microbiol Mol Biol Rev 68:669–685

Hao Y, Winans SC, Glick BR, Charles TC (2010) Identification and characterization of new LuxR/LuxI-type quorum sensing systems from metagenomic libraries. Environ Microbiol 12:105–117

Haque MM, Ghosh TS, Komanduri D, Mande SS (2009) SOrt-ITEMS: Sequence orthology based approach for improved taxonomic estimation of metagenomic sequences. Bioinformatics 25(14):1722–1730

Hardoim PR, van Overbeek LS, van Elsas JD (2008) Properties of bacterial endophytes and their proposed role in plant growth. Trends Microbiol 16:463–471

Hausmann B, Pelikan C, Herbold CW et al (2018) Peatland Acidobacteria with a dissimilatory sulfur metabolism. ISME J 12:1729–1742

Heap IM (1999) International survey of herbicide-resistant weeds: lessons and limitations. In: 1999 Brighton crop protection conference: weeds. Proceedings of an international conference, Brighton, UK, 15-18 November 1999. British Crop Protection Council, Farnham, pp 769–776

Hoerlein G (1994) Glufosinate (phosphinothricin), a natural amino acid with unexpected herbicidal properties. In: Reviews of environmental contamination and toxicology. Springer, New York, pp 73–145

Hoff KJ, Tech M, Lingner T et al (2008) Gene prediction in metagenomic fragments: a large scale machine learning approach. BMC Bioinformatics 9:217

Holben WE, Jansson JK, Chelm BK, Tiedje JM (1988) DNA probe method for the detection of specific microorganisms in the soil bacterial community. Appl Environ Microbiol 54:703–711

Hu J, Okawa H, Yamamoto K et al (2011) Characterization of two cytochrome P450 monooxygenase genes of the pyripyropene biosynthetic gene cluster from *Penicillium coprobium*. J Antibiot (Tokyo) 64:221–227

Huang XF, Chaparro JM, Reardon KF, Zhang R, Shen Q, Vivanco JM (2014) Rhizosphere interactions: root exudates, microbes, and microbial communities. Botany 92(4):267–275

Hughes JB, Hellmann JJ, Ricketts TH, Bohannan BJM (2001) Counting the uncountable: statistical approaches to estimating microbial diversity. Appl Environ Microbiol 67:4399–4406

Hunter PJ, Teakle G, Bending GD (2014) Root traits and microbial community interactions in relation to phosphorus availability and acquisition, with particular reference to *Brassica*. Front Plant Sci 5:27

Huson DH, Auch AF, Qi J, Schuster SC (2007) MEGAN analysis of metagenomic data. Genome Res 17:377–386

Ilangumaran G, Smith DL (2017) Plant growth promoting rhizobacteria in amelioration of salinity stress: a systems biology perspective. Front Plant Sci 8:1768

Ito K, Tanaka T, Hatta R et al (2004) Dissection of the host range of the fungal plant pathogen *Alternaria alternata* by modification of secondary metabolism. Mol Microbiol 52:399–411

Kao-Kniffin J, Carver SM, DiTommaso A (2013) Advancing weed management strategies using metagenomic techniques. Weed Sci 61:171–184

Kennedy AC, Stubbs TL (2007) Management effects on the incidence of jointed goatgrass inhibitory rhizobacteria. Biol Control 40:213–221

Kircher M, Kelso J (2010) High-throughput DNA sequencing–concepts and limitations. BioEssays 32:524–536

Klindworth A, Pruesse E, Schweer T et al (2013) Evaluation of general 16S ribosomal RNA gene PCR primers for classical and next-generation sequencing-based diversity studies. Nucleic Acids Res 41:e1–e1

Kloepper JW, McInroy JA, Liu K, Hu C-H (2013) Symptoms of Fern distortion syndrome resulting from inoculation with opportunistic endophytic fluorescent *Pseudomonas* spp. PLoS One 8: e58531

Konstantinidis KT, Tiedje JM (2007) Prokaryotic taxonomy and phylogeny in the genomic era: advancements and challenges ahead. Curr Opin Microbiol 10:504–509

Kumari A, Kumar R (2018) Exploring Phyllosphere Bacteria for growth promotion and yield of potato (*Solanum tuberosum L.*). Int J Curr Microbiol App Sci 7:1065–1071

Latha PK, Soni R, Khan M et al (2009) Exploration of Csp genes from temperate and glacier soils of the Indian Himalayas and in silico analysis of encoding proteins. Curr Microbiol 58:343–348

Lee HB, Kim C, Kim J et al (2003) A bleaching herbicidal activity of methoxyhygromycin (MHM) produced by an actinomycete strain Streptomyces sp. 8E-12. Lett Appl Microbiol 36:387–391

Lee J-H, Park J-H, Kim J-A et al (2011) Low concentrations of honey reduce biofilm formation, quorum sensing, and virulence in Escherichia coli O157: H7. Biofouling 27:1095–1104

Linning R, Lin D, Lee N et al (2004) Marker-based cloning of the region containing the UhAvr1 avirulence gene from the basidiomycete barley pathogen *Ustilago hordei*. Genetics 166:99–111

Liu N, Li H, Chevrette MG et al (2019) Functional metagenomics reveals abundant polysaccharide-degrading gene clusters and cellobiose utilization pathways within gut microbiota of a wood-feeding higher termite. ISME J 13:104–117

Lydon J, Kong H, Murphy C, Zhang W (2011) The biology and biological activity of *Pseudomonas syringae* pv. tagetis. Pest Technol 5:48–55

Malhotra S, Mishra V, Karmakar S, Sharma RS (2017) Environmental predictors of indole acetic acid producing rhizobacteria at fly ash dumps: nature-based solution for sustainable restoration. Front Environ Sci 5:59

Malmstrom RR, Eloe-Fadrosh EA (2019) Advancing genome-resolved metagenomics beyond the shotgun. MSystems 4:e00118–e00119

Marasco R, Rolli E, Ettoumi B et al (2012) A drought resistance-promoting microbiome is selected by root system under desert farming. PLoS One 7:e48479

Markowitz VM, Chen I-MA, Chu K et al (2012) IMG/M: the integrated metagenome data management and comparative analysis system. Nucleic Acids Res 40:D123–D129

Maruyama T, Ishikura M, Taki H et al (2005) Isolation and characterization of o-xylene oxygenase genes from *Rhodococcus opacus* TKN14. Appl Environ Microbiol 71:7705–7715

Mastouri F, Björkman T, Harman GE (2012) *Trichoderma harzianum* enhances antioxidant defense of tomato seedlings and resistance to water deficit. Mol Plant-Microbe Interact 25:1264–1271

Medd RW, Campbell MA (2005) Grass seed infection following inundation with *Pyrenophora semeniperda*. Biocontrol Sci Tech 15:21–36

Mesapogu S, Babu BK, Bakshi A et al (2011) Rapid detection and quantification of *Fusarium udum* in soil and plant samples using real-time PCR. J Plant Pathol Microbiol 2:2

Mesapogu S, Bakshi A, Babu BK et al (2012) Genetic diversity and pathogenic variability among Indian isolates of *Fusarium udum* infecting pigeonpea (*Cajanus cajan (L.) millsp.*). Int Res J Agric Sci Soil Sci 2:51–57

Miller JR, Koren S, Sutton G (2010) Assembly algorithms for next-generation sequencing data. Genomics 95:315–327

Mirete S, De Figueras CG, González-Pastor JE (2007) Novel nickel resistance genes from the rhizosphere metagenome of plants adapted to acid mine drainage. Appl Environ Microbiol 73:6001–6011

Mirete S, Mora-Ruiz MR, Lamprecht-Grandío M et al (2015) Salt resistance genes revealed by functional metagenomics from brines and moderate-salinity rhizosphere within a hypersaline environment. Front Microbiol 6:1121

Mohanty SR, Dubey G, Ahirwar U et al (2016) Prospect of phyllosphere microbiota: a case study on bioenergy crop Jatropha Curcas. In: Plant-microbe interaction: an approach to sustainable agriculture. Springer, Puchong, pp 453–462

Morgan JAW, Sergeant M, Ellis D et al (2001) Sequence analysis of insecticidal genes from *Xenorhabdus nematophilus PMFI296*. Appl Environ Microbiol 67:2062–2069

Morgante V, Mirete S, de Figueras CG et al (2015) Exploring the diversity of arsenic resistance genes from acid mine drainage microorganisms. Environ Microbiol 17:1910–1925

Müller CA, Obermeier MM, Berg G (2016) Bioprospecting plant-associated microbiomes. J Biotechnol 235:171–180

Nagayama H, Sugawara T, Endo R et al (2015) Isolation of oxygenase genes for indigo-forming activity from an artificially polluted soil metagenome by functional screening using *Pseudomonas putida* strains as hosts. Appl Microbiol Biotechnol 99:4453–4470

Nash MV, Anesio AM, Barker G et al (2018) Metagenomic insights into diazotrophic communities across Arctic glacier forefields. FEMS Microbiol Ecol 94:fiy114

Nath D, Maurya BR, Meena VS (2017) Documentation of five potassium-and phosphorus-solubilizing bacteria for their K and P-solubilization ability from various minerals. Biocatal Agric Biotechnol 10:174–181

Nautiyal CS, Srivastava S, Chauhan PS et al (2013) Plant growth-promoting bacteria *Bacillus amyloliquefaciens NBRISN13* modulates gene expression profile of leaf and rhizosphere community in rice during salt stress. Plant Physiol Biochem 66:1–9

Nocker A, Sossa-Fernandez P, Burr MD, Camper AK (2007) Use of propidium monoazide for live/dead distinction in microbial ecology. Appl Environ Microbiol 73:5111–5117

Ochiai H, Inoue Y, Hasebe A, Kaku H (2001) Construction and characterization of a *Xanthomonas oryzae pv. oryzae* bacterial artificial chromosome library. FEMS Microbiol Lett 200:59–65

Ogram A, Sayler GS, Barkay T (1987) The extraction and purification of microbial DNA from sediments. J Microbiol Methods 7:57–66

Ono A, Miyazaki R, Sota M et al (2007) Isolation and characterization of naphthalene-catabolic genes and plasmids from oil-contaminated soil by using two cultivation-independent approaches. Appl Microbiol Biotechnol 74:501–510

Oulas A, Pavloudi C, Polymenakou P et al (2015) Metagenomics: tools and insights for analyzing next-generation sequencing data derived from biodiversity studies. Bioinform Biol Insights 9:75–88

Pahari A, Mishra BB (2017) Characterization of siderophore producing Rhizobacteria and its effect on growth performance of different vegetables. Int J Curr Microbiol App Sci 6:1398–1405

Panke-Buisse K, Poole AC, Goodrich JK et al (2015) Selection on soil microbiomes reveals reproducible impacts on plant function. ISME J 9:980–989

Peng Y, Leung HCM, Yiu S-M, Chin FYL (2011) Meta-IDBA: a de Novo assembler for metagenomic data. Bioinformatics 27:i94–i101

Pham VHT, Kim J (2012) Cultivation of unculturable soil bacteria. Trends Biotechnol 30:475–484

Ramesh R, Joshi AA, Ghanekar MP (2009) Pseudomonads: major antagonistic endophytic bacteria to suppress bacterial wilt pathogen, *Ralstonia solanacearum* in the eggplant (*Solanum melongena L.*). World J Microbiol Biotechnol 25:47–55

Ray SK, Rajeshwari R, Sonti RV (2000) Mutants of *Xanthomonas oryzae pv. oryzae* deficient in general secretory pathway are virulence deficient and unable to secrete xylanase. Mol Plant-Microbe Interact 13:394–401

Raynaud X, Nunan N (2014) Spatial ecology of bacteria at the microscale in soil. PLoS One 9: e87217

Rees HC, Grant S, Jones B et al (2003) Detecting cellulase and esterase enzyme activities encoded by novel genes present in environmental DNA libraries. Extremophiles 7:415–421

Rodríguez-Segura Z, Chen J, Villalobos FJ et al (2012) The lipopolysaccharide biosynthesis core of the Mexican pathogenic strain *Serratia entomophila* is associated with toxicity to larvae of *Phyllophaga blanchardi*. J Invertebr Pathol 110:24–32

Roesch LFW, Fulthorpe RR, Riva A et al (2007) Pyrosequencing enumerates and contrasts soil microbial diversity. ISME J 1:283–290
Romero M, Martin-Cuadrado A-B, Roca-Rivada A et al (2011) Quorum quenching in cultivable bacteria from dense marine coastal microbial communities. FEMS Microbiol Ecol 75:205–217
Rondon MR, August PR, Bettermann AD et al (2000) Cloning the soil metagenome: a strategy for accessing the genetic and functional diversity of uncultured microorganisms. Appl Environ Microbiol 66:2541–2547
Schauss K, Focks A, Leininger S et al (2009) Dynamics and functional relevance of ammonia-oxidizing archaea in two agricultural soils. Environ Microbiol 11:446–456
Schreiber L, Krimm U, Knoll D et al (2005) Plant–microbe interactions: identification of epiphytic bacteria and their ability to alter leaf surface permeability. New Phytol 166:589–594
Schröder C, Elleuche S, Blank S, Antranikian G (2014) Characterization of a heat-active archaeal β-glucosidase from a hydrothermal spring metagenome. Enzym Microb Technol 57:48–54
Sessitsch A, Hardoim P, Döring J et al (2012) Functional characteristics of an endophyte community colonizing rice roots as revealed by metagenomic analysis. Mol Plant-Microbe Interact 25:28–36
Shan W, Hardham AR (2004) Construction of a bacterial artificial chromosome library, determination of genome size, and characterization of an Hsp70 gene family in *Phytophthora nicotianae*. Fungal Genet Biol 41:369–380
Silva CC, Hayden H, Sawbridge T et al (2013) Identification of genes and pathways related to phenol degradation in metagenomic libraries from petroleum refinery wastewater. PLoS One 8: e61811
Smith CJ, Osborn AM (2009) Advantages and limitations of quantitative PCR (Q-PCR)-based approaches in microbial ecology. FEMS Microbiol Ecol 67:6–20
Smith RG (2015) A succession-energy framework for reducing non-target impacts of annual crop production. Agric Syst 133:14–21
Song Y (2019) A new CRISPR scissor. Nat Chem Biol 15:315
Sørensen J, Nicolaisen MH, Ron E, Simonet P (2009) Molecular tools in rhizosphere microbiology—from single-cell to whole-community analysis. Plant Soil 321:483–512
Strange RN (2007) Phytotoxins produced by microbial plant pathogens. Nat Prod Rep 24:127–144
Streit WR, Schmitz RA (2004) Metagenomics–the key to the uncultured microbes. Curr Opin Microbiol 7:492–498
Suenaga H, Ohnuki T, Miyazaki K (2007) Functional screening of a metagenomic library for genes involved in microbial degradation of aromatic compounds. Environ Microbiol 9:2289–2297
Sulaiman S, Yamato S, Kanaya E et al (2012) Isolation of a novel cutinase homolog with polyethylene terephthalate-degrading activity from leaf-branch compost by using a metagenomic approach. Appl Environ Microbiol 78:1556–1562
Tan H, Mooij MJ, Barret M et al (2014) Identification of novel phytase genes from an agricultural soil-derived metagenome. J Microbiol Biotechnol 24:113–118
Tank N, Saraf M (2010) Salinity-resistant plant growth promoting rhizobacteria ameliorates sodium chloride stress on tomato plants. J Plant Interact 5:51–58
Tchigvintsev A, Tran H, Popovic A et al (2015) The environment shapes microbial enzymes: five cold-active and salt-resistant carboxylesterases from marine metagenomes. Appl Microbiol Biotechnol 99:2165–2178
Timmusk S, El-Daim IAA, Copolovici L et al (2014) Drought-tolerance of wheat improved by rhizosphere bacteria from harsh environments: enhanced biomass production and reduced emissions of stress volatiles. PLoS One 9:e96086
Torsvik V, Goksøyr J, Daae FL (1990) High diversity in DNA of soil bacteria. Appl Environ Microbiol 56:782–787
Trognitz F, Hackl E, Widhalm S, Sessitsch A (2016) The role of plant–microbiome interactions in weed establishment and control. FEMS Microbiol Ecol 92:fiw138
Tyson GW, Chapman J, Hugenholtz P et al (2004) Community structure and metabolism through reconstruction of microbial genomes from the environment. Nature 428:37–43

Unno Y, Shinano T (2012) Metagenomic analysis of the rhizosphere soil microbiome with respect to phytic acid utilization. Microbes Environ 28(1):120–127

Van Wees SCM, Van der Ent S, Pieterse CMJ (2008) Plant immune responses triggered by beneficial microbes. Curr Opin Plant Biol 11:443–448

Vessey JK (2003) Plant growth promoting rhizobacteria as biofertilizers. Plant Soil 255:571–586

Wang C-J, Yang W, Wang C et al (2012) Induction of drought tolerance in cucumber plants by a consortium of three plant growth-promoting rhizobacterium strains. PLoS One 7:e52565

Weissmann R, Uggla C, Gerhardson B (2003) Field performance of a weed-suppressing *Serratia plymuthica* strain applied with conventional spraying equipment. BioControl 48:725–742

Whisson S, Lee T, Bryan G et al (2001) Physical mapping across an avirulence locus of *Phytophthora infestans* using a highly representative, large-insert bacterial artificial chromosome library. Mol Gen Genomics 266:289–295

White JR, Nagarajan N, Pop M (2009) Statistical methods for detecting differentially abundant features in clinical metagenomic samples. PLoS Comput Biol 5:e1000352

Williamson LL, Borlee BR, Schloss PD et al (2005) Intracellular screen to identify metagenomic clones that induce or inhibit a quorum-sensing biosensor. Appl Environ Microbiol 71:6335–6344

Yang J, Kloepper JW, Ryu C-M (2009) Rhizosphere bacteria help plants tolerate abiotic stress. Trends Plant Sci 14:1–4

Yao Q, Li Z, Song Y et al (2018) Community proteogenomics reveals the systemic impact of phosphorus availability on microbial functions in tropical soil. Nat Ecol Evol 2:499–509

Zdor RE, Alexander CM, Kremer RJ (2005) Weed suppression by deleterious rhizobacteria is affected by formulation and soil properties. Commun Soil Sci Plant Anal 36:1289–1299

12 The Skin Metagenomes: Insights into Involvement of Microbes in Diseases

Jyotsana Sharma, Varun Sharma, and Indu Sharma

Abstract

The skin is the first barrier of the body and is also a pitch for various microbes. Earlier, research on the skin microbiology was primarily influenced by culture-based techniques. However, now the scenario has changed. Metagenomics has emerged as a more efficient tool for obtaining comprehensive information and understanding of the microbiome of the skin. Metagenomics comprises techniques of high-throughput sequencing, gene-prediction, sequencing of amplicon-based assays, metatranscriptomics and shotgun metagenomics. The sequence data is studied further using statistical as well as comparative analyses. These analyses provide an understanding of the skin microbiome diversity, both functionally and metabolically. Thus, better information of the mechanism of how microbial population of the skin interacts with one other and also with the host could lead to the development of significant clinical treatments as to how we can control those microbial interactions for preventing as well as in curing of various dermatological diseases.

Keywords

Culture-independent techniques · Dermatological disorders · Metagenomics · Microbiome · Skin

J. Sharma
School of Biotechnology, Shri Mata Vaishno Devi University, Katra, Jammu and Kashmir, India

V. Sharma · I. Sharma (✉)
Ancient DNA Laboratory, Birbal Sahni Institute of Palaeosciences, Lucknow, Uttar Pradesh, India

R. S. Chopra et al. (eds.), *Metagenomics: Techniques, Applications, Challenges and Opportunities*, https://doi.org/10.1007/978-981-15-6529-8_12

12.1 Introduction

Skin is the primary outermost layer that acts as a barrier to the pathogens and protects the body from these foreign microorganisms and other toxins. Moreover, it also harbors diverse microorganisms on it that includes bacteria (symbiotic or commensal), microeukaryotes, archaea and viruses (Grice 2015). This composite ecosystem of microorganisms of skin consists of moist areas that cover finger web spaces, armpit, sebaceous environments including face as well as back, hard parts including the forearm, buttock, areas having hair follicles and skin folds. The diverse community of microorganisms which includes fungi, bacteria, viruses, parasites play a considerable part in the development of dermatological disorders (Mathieu et al. 2013; Wilantho et al. 2017). There are numerous reports in which the culture-dependent approaches were used for studying the skin microorganisms. However, these studies were less proficient because the percentage of the bacterial species that can be cultivated comes out to be lesser than 1% due to which bulk of microorganisms remained unidentified. Therefore, techniques like metagenomics and high-throughput sequencing are used in order to perform an unbiased identification as well as characterization of the skin microbiome (Mende et al. 2012; Chen and Tsao 2013).

Metagenome allows the involvement of all the genes, as well as genetic elements within the microorganisms and the host. The term Metagenomics is defined as the study of diverse communities of microbes at different levels, i.e. structural as well as functional, followed by their interactions with the hosts as well (Martín et al. 2014). In this article, we aim to present the methods used for the study and characterization of the skin microbiome and the role of metagenomics in an extensive understanding of the skin disorders and also the health of the skin. Metagenomics involves gene-amplification, DNA sequencing, and study of the Hyper Variable Regions (HVR) of the bacterial 16S ribosomal RNA gene among other phylogenetic marker genes (Mathieu et al. 2013; Martín et al. 2014). The study by metagenomics includes techniques like metagenomic DNA extraction, construction of metagenomic libraries, taxonomic composition analysis, whole-genome sequencing, and statistical analysis. All this advancement has upgraded our acquaintance regarding the microbiome of skin and the interactions it is having with the host and its contribution in different dermatological disorders which has led to the development of treatments for these diseases via diagnostic, prognostic and various therapeutic means (Kergourlay et al. 2015). Earlier, there were several myths regarding the prevalence of diseases like dandruff and acne that they were caused because of the existence of bacterium *Malassezia fungi* and *Propionibacterium acnes,* respectively. However, some metagenomic studies have reported the contribution of complex microbial communities in the development of these diseases. For that reason, modern techniques allow for direct identification of microbes in the sample but do not differentiate between dead and alive species. (dandruff, seborrheic dermatitis, psoriasis, bovine digital dermatitis, acne vulgaris, basal cell carcinoma, melanoma, erythema, and atopic dermatitis, among others) have been studied, analyzed and investigated through these metagenomic approaches (Horton et al. 2015; Guet-Revillet et al. 2017).

12.2 Brief Layout of Skin Ecosystem

Due to the phenomenon of epidermis cohesion, skin serves as a physical blockade to the infections and as a result protects our body from foreign organisms. Moreover, it has been observed that a subtle balance between host and the skin microbiota exists and skin disorders are developed only due to the disruptions in this balance (Hannigan and Grice 2013; Segre 2006). These dermatological conditions can be well-studied by analyzing the skin microbiome and also its interactions with the host (Capone et al. 2011). The environment of the skin varies, for instance; it depends upon certain factors like thickness of skin, presence of glands, hair follicles density, among others. Different glands harbor different microenvironments e.g. sweat glands which includes apocrine and eccrine glands where eccrine glands constantly bath/wash the surface of the skin with sweat which mostly contain salts and water. On the other hand, hair follicles and sebaceous glands present their distinctive microbial population where sebaceous glands exude sebum that guards, lubricates and smears the skin which somehow acts an antibacterial safeguard for it. Based upon the contour of the skin, it has been observed that there exist few patches that are moderately occluded having higher temperature as well as humidity (groin, and toe web), in comparison few areas that are extremely desiccated like arms and legs which are exposed to huge fluctuations in temperature. Additionally, areas like face, back and chest that are endowed with sebaceous glands. The physiology of the Skin helps in determining the prototype of the microbial population that inhabits the skin environment (Grice and Segre 2011; Proksch et al. 2008). The most abundant species of microorganisms that are found on the moist sites are Staphylococcus and Corynebacterium species whereas *Propionibacterium* and *Malassezia* species are abundant in lipophilic environment. Mixed population of microorganisms can be found in the dry areas that includes including Bacteroidetes, Actinobacteria, Firmicutes and Proteobacteria. Most of the bacteria found on skin can be classified into: Actinobacteria, Bacteroidetes, Firmicutes and Proteobacteria. Similar phyla are also present on the site of inner mucosal surfaces but in contrasting to the gut that is densely populated with Bacteroidetes and Firmicutes, skin surface is dominated by the Actinobacteria (Gao et al. 2007; Grice et al. 2009). On Comparing, gut and skin, it can be concluded that the nutrition wise skin is extremely poor as sources of nutrients are only sweat and sebum. Sweat is composed of urea, lactic acid, amino acids and proteins whereas triglycerides, fatty acids, squalene and wax esters constituents of the human sebum (Elias 2007). One of the most widespread microbes found on the skin is Staphylococcus epidermidis which is a gram-positive strain and is a Firmicutes. These are very much adjustable in occluded and moist skin sites however they are also adaptable in drier and more exposed areas. *Corynebacterium* which is Gram-positive, grows on sebaceous or moist skin surfaces. These are lipohilic but make use of the lipid contents from stratum corneum and sebum.

Furthermore, generally *Corynebacterium* grows in the sites having high salt concentrations which means they can be found on the areas affluent with eccrine glands. Propionibacteria are gram positive and mainly localized in hair follicles, which means sebum rich sites of the skin. *Malassezia* genus is the major fungus that is found on the surface of the skin which plays a pivotal part in the development of frequent skin diseases like dandruff (Tanaka et al. 2016; Cogen et al. 2008). With the advent of the technique of Whole-genome shotgun metagenomics it has become achievable to investigate and study various skin viruses like human papillomavirus and human polyomaviruses efficiently (Arroyo Mühr et al. 2015; Ma et al. 2014).

12.3 Novel Approaches Based on Culture-Independent Techniques and Development of Personalized Treatment

Previously the culturing methods were used to acquire knowledge regarding the skin-associated microbes by phylogenetic and taxonomic profiling via phenotypic, microscopic and biochemical associations. Although bulks of microorganisms are not capable of growing under precise conditions, hence this mode of study extensively ignored the mixed microbial population (Hannigan and Grice 2013). Consequently, the introduction of metagenomics increased the research interest and also led to the treatment of a range of dermatological disorders. The study of metagenomics involves a few necessary steps that are shown in Fig. 12.1. These steps are briefly elaborated as follows: isolating DNA from a sample followed by cloning into an appropriate vector followed by transformation of a host bacterial strain. The next step is the screening of the transformants for 16S rRNA and recA genes. Screening may also be performed for expression of specific traits or the discovery of more conserved genes (Lau et al. 2017). Next in the metagenomic analysis of the microbiomes is the generation of amplicons from the pooled library clones, followed by adapter-mediated sequencing. The sequencing results are then analyzed and assembled using computational methods, statistical analysis and further comparison based studies (Kim et al. 2017). The method of shotgun metagenomic involves collection and investigation of the total amount of DNA recovered from the community devoid of depending directly on marker genes and rather, allows for sequencing and analysis of the full genetic potential of the given sample (NRC 2007). Ribosomal community profiling is an additional technique in which the ribosomal RNA marker gene is used. It includes those conserved regions that permit PCR primer binding as well as phylogenetic analysis together with variable regions also the sequences of which prove to be very useful in deducing the species-richness and taxonomic diversity of the microbial community (Grice 2015). Metatranscriptomics is also a valuable method to analyze the species that are profusely present rather than obtaining DNA/RNA from a sample of skin which is then further sequenced using next-generation sequencing (NRC 2007; Schuster 2008). Cloning of the DNA before sequencing is not necessarily required in case of high-throughput methods which makes the whole procedure comparatively less exhausting. The precision of the obtained assemblies can be enhanced by rectifying

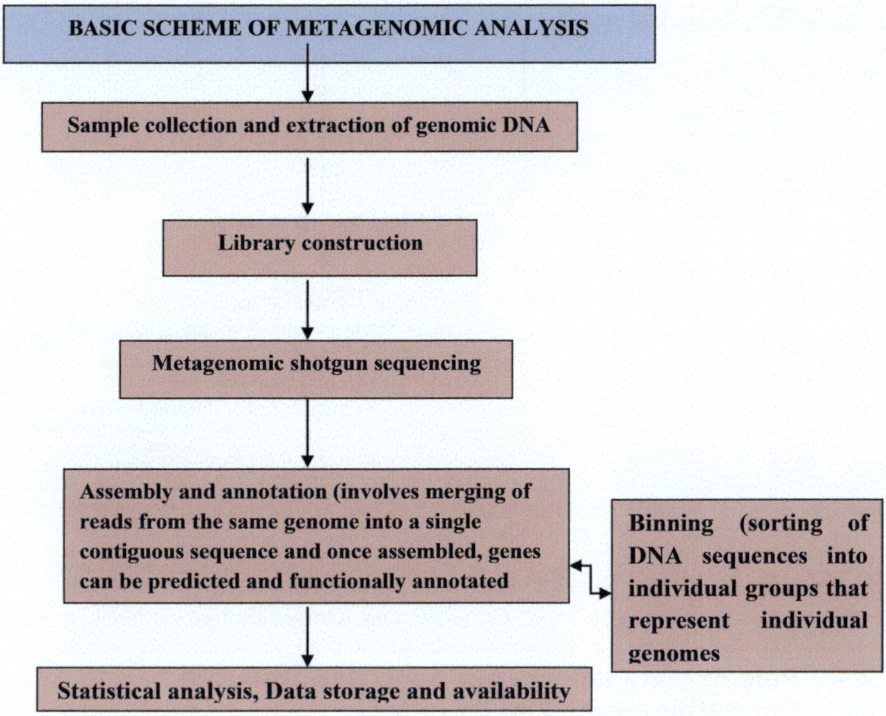

Fig. 12.1 Schematic representation of taxonomy identification for metagenomic high throughput sequencing analysis

misassemblies by using paired-end tags provided by a range of assembly programs e.g. Phrap and velvet assembler (Chen and Pachter 2005). Another tool BLAST is used for quick investigation of phylogenetic markers (Wooley et al. 2010). PhymmBL, AMPHORA, and SLIMM are the tools that are used during binning of the sequences. Binning means an association of a particular sequence with a species, in order to compare the diversity. Binning is followed by assigning them to a particular operational taxonomic unit (OTU) (Kunin et al. 2008). Tools like CLARK can carry out this taxonomic annotation at tremendously high pace than that of MG-RAST or MEGAN, which are BLAST-based methods (Segata et al. 2012). Hence, metagenomic approaches are critical to the prospecting of useful sequence information from big-datasets which leads to the scrutiny of the functions of the skin microbiome. There is also a potential for information about genes that encodes virulence and pathogenicity and therefore can prove very helpful in developing novel therapeutics in order to cure such skin related diseases.

Development of personalized medicines and various probiotic-based treatments in case of skin diseases are one of the significant breakthroughs of the metagenomics (Fig. 12.2).

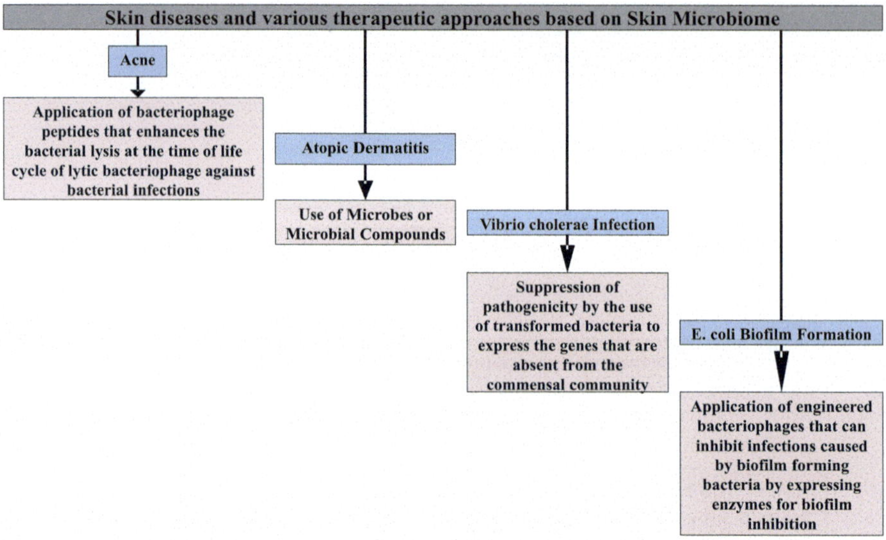

Fig. 12.2 Skin diseases and various therapeutic approaches based on skin microbiome

12.4 Role of Metagenomics in Analyzing Different Dermatological (Skin) Disorders

12.4.1 Psoriasis

It is an inflammatory disease of the skin that affects 2–3% population of the whole of the world. The most prevalent type of psoriasis that is found in 85–90% of affected patients is Plaque psoriasis. Severity in this skin disease could also lead to the development of certain other significant diseases like cardiovascular, psoriatic arthritis and metabolic syndromes (Rapp et al. 1999; Grozdev et al. 2014). Studies performed by culture-based techniques recognized that *Malassezia*, *S. aureus*, beta-hemolytic streptococci, and *Enterococcus faecalis* are the contributors of psoriasis. However, on compositional analysis using the 16S rRNA marker, it was found that dysregulation in the skin microbiota due to the neonatal antibiotic treatment led to the growth of psoriasis. With High-resolution shotgun metagenomics, it has been observed that psoriasis is developed due to the decrease in microbial diversity and with an increase in Staphylococcus strain-level variability (Tett et al. 2017; Zanvit et al. 2015). Studies based on metagenomics are highly appreciated in having information about taxonomic differences of the microbial population involved in psoriasis and therefore presents a considerable substitute to the studies that are culture-based.

12.4.2 Acne

Acne is common, occurring condition of the skin that is characterized by deformity in the production of the sebum by the hair follicle, which is also known as a pilosebaceous unit. It more commonly affects adolescents in comparison to adults (White 1998; Tan 2003; Barnard et al. 2016). 16S rRNA-based metagenomic study reported that Propionibacterium acnes account for almost 90% of the microbial environment of the skin together with *Propionibacterium humerusii*, *Propionibacterium granulosum* and *Staphylococcus epidermidis* (Fitz-Gibbon et al. 2013). It has also been reported that such skin diseases are not due to the entire species being pathogenic; instead, some particular strains of that species cause these diseases. For example, *P. acnes* also maintain the healthiness of the skin health by preventing opportunistic pathogens by maintaining the acidic pH of the skin (Liu et al. 2015). Metagenomics helps in the determination of microbial interactions as it offers more coverage than the traditional approaches. When the microbiota from the patients having acne was compared to the healthy individuals, there was not any noteworthy distinction found in the abundance of *P. acnes*. The assessment of *P. acnes* at the strain level by defining every unique sequence of 16S rDNA as a 16S rDNA allele type permits us to contrast the populations of the strain of *P. Acnes* (Wilantho et al. 2017; Barnard et al. 2016). The equilibrium between acne and metagenomic constituents shows the virulence, health properties and function of the skin microbiota in the development of certain diseases of the skin (Kwon and Suh 2016). Therefore, it provides insights into the skin microbiota and also the possible mechanism of pathogenesis in acne. The information about these mechanisms can lead to designing of certain probiotics as well as phage therapies for treating acne and thus also help in maintaining the healthiness of the skin.

12.4.3 Atopic Dermatitis

Atopic dermatitis (AD) is also an inflammatory disease of the skin but is non-infectious and is majorly found in children. Patients with this disease have transformed microbial community which is mainly coupled with the occurrence of Staphylococcus aureus (Ong et al. 2002). A metagenomic study based on 16S-rRNA of AD has revealed that the population *S. epidermidis* and *S. aureus* increases. In turn, there are changes in the availability of native non-staphylococcal populations resulting in a decrease of the bacterial diversity, including Alcaligenaceae, Sediminibacterium and *Lactococcus* (Kim et al. 2017). Therefore, the metagenomic analysis is important for studying the mechanism of invasion and colonization of these pathogens.

12.4.4 Dandruff and Seborrheic Dermatitis

Another inflammatory state of the scalp is known as Dandruff. Dandruff is commonly associated with problems like severe itching along with scaling of the skin (Soares et al. 2016). The more chronic form of dandruff is termed as Seborrheic dermatitis that also spreads to the other areas of the body that are having sebaceous glands other than the scalp. Earlier researches reported that it is mainly caused by the fungi *Malassezia* (Tanaka et al. 2016; Dawson Jr 2007; DeAngelis et al. 2005). However, later reports suggested that it is a collective contribution of mixed microbial populations (Byrd et al. 2017). Other populations include Propionibacterium, Staphylococcus and Corynebacterium along with the *Malassezia* sp. as have been found by Next-generation sequencing performed on patients having dandruff and healthy individuals. Numerous Metagenomic and molecular studies by 26S r RNA molecular analysis have revealed that concentration of Propionibacterium acnes is higher in normal scalps while in case of patients with dandruff, microbial population that dominates includes *Staphylococcus epidermidis, Leptotrichia, Pseudomonas, Micrococcus, Erwinia, Selenomonas, Enhydrobacter*, fungal genera including *Malassezia*, *Candida, Filobasidium* and *Aspergillus*. These scientific advancements led to increase in our understanding of the role of the microbiome, etiology of the diseases, symptom development which could be very beneficial in development of the therapeutic procedures needed to combat these diseases (Wan et al. 2017).

12.5 Conclusion

Metagenomics revolutionize the whole area of diagnosing the causes of the skin diseases to its depth, but the challenge is right application of this information to build up certain diagnostic and therapeutic means for the cure of these skin diseases. Improved understanding of the microbial environment of the skin will help in development of certain probiotic and prebiotic treatments considering rising antibiotic resistance over medically significant bacterial populations. So, there will probably be an expanded enthusiasm for other substitute ways to deal with treating skin infections and also slow down the pace of the spreading antibiotic resistance. So, here Metagenomics could play a promising role in development of treatments and precision medicines that could prove very beneficial in management, prevention and treatment of the chronic skin disorders.

References

Arroyo Mühr LS, Hultin E, Bzhalava D et al (2015) Human papillomavirus type 197 is commonly present in skin tumors. Int J Cancer 136:2546–2555

Barnard E, Shi B, Kang D et al (2016) The balance of metagenomic elements shapes the skin microbiome in acne and health. Sci Rep 6:39491

Byrd AL, Deming C, Cassidy SKB et al (2017) Staphylococcus aureus and Staphylococcus epidermidis strain diversity underlying pediatric atopic dermatitis. Sci Transl Med 9:eaal4651

Capone KA, Dowd SE, Stamatas GN, Nikolovski J (2011) Diversity of the human skin microbiome early in life. J Invest Dermatol 131:2026–2032
Chen K, Pachter L (2005) Bioinformatics for whole-genome shotgun sequencing of microbial communities. PLoS Comput Biol 1:e24–e112
Chen YE, Tsao H (2013) The skin microbiome: current perspectives and future challenges. J Am Acad Dermatol 69:143–155
Cogen AL, Nizet V, Gallo RL (2008) Skin microbiota: a source of disease or defence? Br J Dermatol 158:442–455
Council NR (2007) The new science of metagenomics: revealing the secrets of our microbial planet. National Academies Press, Washington, D.C.
Dawson TL Jr (2007) Malassezia globosa and restricta: breakthrough understanding of the etiology and treatment of dandruff and seborrheic dermatitis through whole-genome analysis. J Investig Dermatol Symp Proc 12:15–19
DeAngelis YM, Gemmer CM, Kaczvinsky JR et al (2005) Three etiologic facets of dandruff and seborrheic dermatitis: Malassezia fungi, sebaceous lipids, and individual sensitivity. J Investig Dermatol Symp Proc 10(3):295–297
Elias PM (2007) The skin barrier as an innate immune element. Semin Immunopathol 29(1):3–14
Fitz-Gibbon S, Tomida S, Chiu B-H et al (2013) Propionibacterium acnes strain populations in the human skin microbiome associated with acne. J Invest Dermatol 133:2152–2160
Gao Z, Tseng C, Pei Z, Blaser MJ (2007) Molecular analysis of human forearm superficial skin bacterial biota. Proc Natl Acad Sci 104:2927–2932
Grice EA (2015) The intersection of microbiome and host at the skin interface: genomic-and metagenomic-based insights. Genome Res 25:1514–1520
Grice EA, Kong HH, Conlan S et al (2009) Topographical and temporal diversity of the human skin microbiome. Science (80-) 324:1190–1192
Grice EA, Segre JA (2011) The skin microbiome. Nat Rev Microbiol 9:244–253
Grozdev I, Korman N, Tsankov N (2014) Psoriasis as a systemic disease. Clin Dermatol 32:343–350
Guet-Revillet H, Jais J-P, Ungeheuer M-N et al (2017) The microbiological landscape of anaerobic infections in hidradenitis suppurativa: a prospective metagenomic study. Clin Infect Dis 65:282–291
Hannigan GD, Grice EA (2013) Microbial ecology of the skin in the era of metagenomics and molecular microbiology. Cold Spring Harb Perspect Med 3:a015362
Horton JM, Gao Z, Sullivan DM et al (2015) The cutaneous microbiome in outpatients presenting with acute skin abscesses. J Infect Dis 211:1895–1904
Kergourlay G, Taminiau B, Daube G, Vergès M-CC (2015) Metagenomic insights into the dynamics of microbial communities in food. Int J Food Microbiol 213:31–39
Kim M-H, Rho M, Choi J-P et al (2017) A metagenomic analysis provides a culture-independent pathogen detection for atopic dermatitis. Allergy Asthma Immunol Res 9:453–461
Kunin V, Copeland A, Lapidus A et al (2008) A bioinformatician's guide to metagenomics. Microbiol Mol Biol Rev 72:557–578
Kwon HH, Suh DH (2016) Recent progress in the research about Propionibacterium acnes strain diversity and acne: pathogen or bystander? Int J Dermatol 55:1196–1204
Lau P, Cordey S, Brito F et al (2017) Metagenomics analysis of red blood cell and fresh-frozen plasma units. Transfusion 57:1787–1800
Liu J, Yan R, Zhong Q et al (2015) The diversity and host interactions of Propionibacterium acnes bacteriophages on human skin. ISME J 9:2078–2093
Ma Y, Madupu R, Karaoz U et al (2014) Human papillomavirus community in healthy persons, defined by metagenomics analysis of human microbiome project shotgun sequencing data sets. J Virol 88:4786–4797
Martín R, Miquel S, Langella P, Bermúdez-Humarán LG (2014) The role of metagenomics in understanding the human microbiome in health and disease. Virulence 5:413–423

Mathieu A, Delmont TO, Vogel TM et al (2013) Life on human surfaces: skin metagenomics. PLoS One 8:e65288

Mende DR, Waller AS, Sunagawa S et al (2012) Assessment of metagenomic assembly using simulated next generation sequencing data. PLoS One 7(2):e31386

Ong PY, Ohtake T, Brandt C et al (2002) Endogenous antimicrobial peptides and skin infections in atopic dermatitis. N Engl J Med 347:1151–1160

Proksch E, Brandner JM, Jensen J (2008) The skin: an indispensable barrier. Exp Dermatol 17:1063–1072

Rapp SR, Feldman SR, Exum ML et al (1999) Psoriasis causes as much disability as other major medical diseases. J Am Acad Dermatol 41:401–407

Schuster SC (2008) Next-generation sequencing transforms today's biology. Nat Methods 5:16–18

Segata N, Waldron L, Ballarini A et al (2012) Metagenomic microbial community profiling using unique clade-specific marker genes. Nat Methods 9:811–814

Segre JA (2006) Epidermal barrier formation and recovery in skin disorders. J Clin Invest 116:1150–1158

Soares RC, Camargo-Penna PH, de Moraes V et al (2016) Dysbiotic bacterial and fungal communities not restricted to clinically affected skin sites in dandruff. Front Cell Infect Microbiol 6:157

Tan H-H (2003) Antibacterial therapy for acne. Am J Clin Dermatol 4:307–314

Tanaka A, Cho O, Saito C et al (2016) Comprehensive pyrosequencing analysis of the bacterial microbiota of the skin of patients with seborrheic dermatitis. Microbiol Immunol 60:521–526

Tett A, Pasolli E, Farina S et al (2017) Unexplored diversity and strain-level structure of the skin microbiome associated with psoriasis. NPJ Biofilms Microbiomes 3:1–12

Wan T-W, Higuchi W, Khokhlova OE et al (2017) Genomic comparison between Staphylococcus aureus GN strains clinically isolated from a familial infection case: IS1272 transposition through a novel inverted repeat-replacing mechanism. PLoS One 12:e0187288

White GM (1998) Recent findings in the epidemiologic evidence, classification, and subtypes of acne vulgaris. J Am Acad Dermatol 39:S34–S37

Wilantho A, Deekaew P, Srisuttiyakorn C et al (2017) Diversity of bacterial communities on the facial skin of different age-group Thai males. PeerJ 5:e4084

Wooley JC, Godzik A, Friedberg I (2010) A primer on metagenomics. PLoS Comput Biol 6: e1000667

Zanvit P, Konkel JE, Jiao X et al (2015) Antibiotics in neonatal life increase murine susceptibility to experimental psoriasis. Nat Commun 6:1–10

Computational Metagenomics: State-of-the-Art, Facts and Artifacts

13

Harpreet Singh, Purnima Sharma, Rupinder Preet Kaur, Diksha Thakur, and Pardeep Kaur

Abstract

Microbes influence almost every process of life in one or the other way. We have been employing various techniques to study microbial communities and the ways they interact with the environment. With advances in molecular biology, we are now able to study microbes which were unculturable a few decades ago. The study of microbial communities by sequencing DNA samples from various ecological niches is called metagenomics. Parallel advancements in whole-genome sequencing (WGS) technologies have enabled the scientific community to explore complex biological samples cost-effectively. Generation of billions of reads in a single NGS run poses a significant challenge to store, manage and analyze this vast data. Development of novel bioinformatics applications, specifically for metagenomics data, is therefore vital to assign biological significance to metagenomics data. Many strategies have been introduced to deal with quality control, assembly, binning and functional annotation of metagenomic data. However, there is a great scope of improvement in terms of the diversity, uneven representation, variability and scale of data. Nowadays, the focus has also been shifted to integrate various software tools in a logical manner to develop dedicated pipelines built on easy to use graphic-rich interface. So, continuous improvement in computational resources, as well as bioinformatics software applications, is required in order to capture real information from the mammoth of data being generated in metagenome sequencing experiments.

H. Singh (✉) · P. Sharma · D. Thakur · P. Kaur
Department of Bioinformatics, Hans Raj Mahila Maha Vidyalaya, Jalandhar, Punjab, India

R. P. Kaur
Department of Chemistry, Guru Nanak Dev University College Verka, Amritsar, Punjab, India
e-mail: rupinder_chemverka@gndu.ac.in

R. S. Chopra et al. (eds.), *Metagenomics: Techniques, Applications, Challenges and Opportunities*, https://doi.org/10.1007/978-981-15-6529-8_13

13.1 Introduction

Microbes are essential in the ecosystem and influence every aspect of life on this mother Earth in one way or the other. The number of ways in which microbes influence daily life is countless. All plants and animals have microbial communities as their inherent partners who help them with the availability of essential nutrients, metals, and vitamins. The microbial communities hosted by plants or living around them play a crucial role in maintaining their health and productivity. Additionally, microbes have been useful to remediate toxins in the environment, both natural as well as produced by human activities including oil and chemical spills. They have evolved to be a part of diverse strata of this biosphere including soil, water and air, apart from being co-evolving within their living hosts including gut of higher organisms.

The microbes present in the human intestine and mouth helps to extract energy from food that we could not digest, apart for providing us protection against a variety of disease-causing pathogens. The recent surge in the antibiotic resistance of infectious pathogens makes it more essential for us to understand the role of microbial communities in the development of said resistance. The detailed understanding of mechanisms by which microbial communities can aid our protection from infectious agents may enable us to exploit specific genes or mechanisms to develop future medicines.

Our understanding of life at the molecular level gave us new opportunities to explore the microbial world using advanced molecular biological techniques. In the late 1970s, Carl Woese gave an idea of using ribosomal RNA sequences as markers to classify living organisms (Woese and Fox 1977). Fortunately, the introduction of Sanger automated sequencing at the same time, made it possible to sequence ribosomal RNA genes from several species. Both these developments revolutionized the way in which microorganisms are studied and classified. Due to established evidence that many microbes resist being cultured, the use of molecular markers provided a successful alternative identifying and enumerating microbes without culturing them in laboratories (Pace 1997). Later on, further advances in molecular biology techniques such as polymerase chain reaction (PCR), rRNA gene cloning and sequencing, fluorescent in situ hybridization (FISH), denaturing gradient gel electrophoresis (DGGE), thermal gradient gel electrophoresis (TGGE), restriction-fragment length polymorphism (RFLP), and terminal restriction fragment length polymorphism (T-RFLP) accelerated the exploration and understanding of microbial diversity and provided insights into a "new uncultured world" of microbial communities (Escobar-Zepeda et al. 2015). Despite such advancements in studying the microbial world, many aspects of microbial life remain unanswered, especially about their metabolic and ecological functions. The journey continued with motivation from efforts to discover new genes, explore new functions, and to extract novel metabolic products. These increasing technological applications gave birth to biotechnology. Some successful discoveries include "terragines" from *Streptomyces lividians* (Wang et al. 2000) or genes for production of broad-spectrum antibiotics from soil-DNA libraries (Gillespie et al. 2002). These discoveries set the stage for a new specialization called "metagenomic analysis," which was later stated as the collection of all genomes from microorganisms in an ecosystem (Handelsman et al. 1998).

13.1.1 Metagenomics: A New Way to Explore Microbial Communities

Metagenomics is a powerful combination of genomics, bioinformatics, and systems biology which can be used to study the genomes of many organisms simultaneously. In fact, 'meta', in Greek means "beyond," and 'genomics' mean the study of the entire DNA content of an organism (Handelsman et al. 1998). The metagenomics, therefore, simply means the study of many genomes at a time. Metagenomics provides new access to the microbial world by making it possible to study uncultured microbial communities which were otherwise impossible to explore. It provides access to the untapped reservoir of novel enzymes, metabolites and other chemicals. Metagenomic approaches can be used as a powerful tool to directly isolate nucleic acids from environmental samples to compare and explore the ecology (Biddle et al. 2008), metabolic profiling (DeLong et al. 2006; Tringe et al. 2005) and identifying novel biomolecules (Daniel 2005; Ferrer et al. 2009; Handelsman 2004; Simon and Daniel 2010; Steele et al. 2009).

Parallel advancements in whole-genome sequencing (WGS) technologies have given us several opportunities to explore complex samples from biomedical as well as environmental experiments at the molecular level. Significant milestones of metagenomics have been summarized in Fig. 13.1. It is now possible to sequence billions of reads in a single NGS run. Also, the generation of novel and robust bioinformatics software programs and pipelines has further revolutionized the analysis of single genes and proteins to a collection of entire sets of molecules from whole genomes. Furthermore, recent innovations and improvements in sequencing instruments has led to an exponential decrease in sequencing costs, accelerating the adoption of NGS technologies by various stakeholders (Chiu and Miller 2019). The excessive use of NGS in metagenomics generates an enormous amount of raw data to be processed and analyzed, making bioinformatics analysis a significant bottleneck (Scholz et al. 2012).

We, therefore, need to develop bioinformatics applications to deal with metagenomic data specifically, otherwise, there will be a massive gap between the data being generated and its biological significance (Escobar-Zepeda et al. 2015). We need to develop novel strategies to tackle challenges in quality control, assembly, binning and functional annotation of metagenomic data in terms of its diversity, uneven representation, variability and scale. Nowadays, the focus has also been shifted to integrate various software tools logically to develop dedicated pipelines built on easy to use graphic-rich interfaces (Jünemann et al. 2017). Besides, real-world metagenomics datasets require huge storage and computational resources giving rise to cloud-based solutions (Wilkening et al. 2009).

We can summarize that the science of metagenomics has a great potential to explore new avenues for discovering novel genes, enzymes and chemicals for

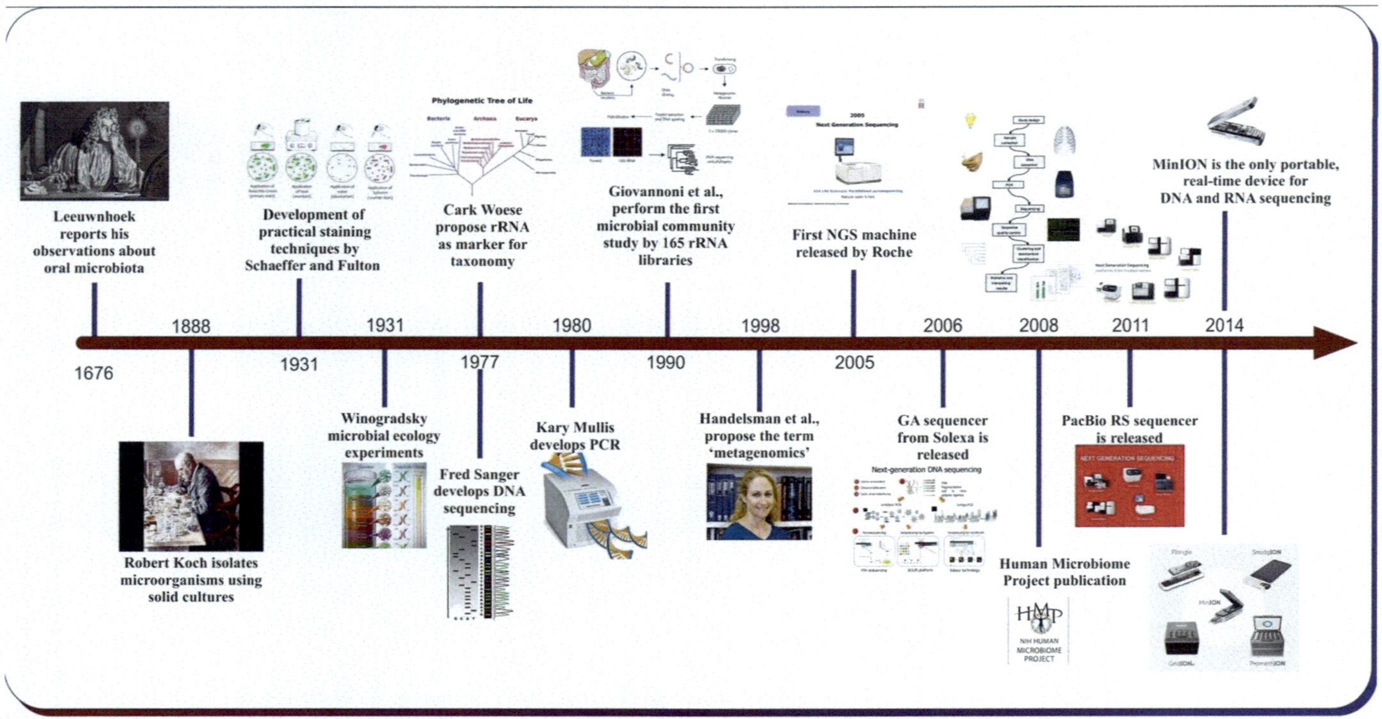

Fig. 13.1 Timeline of major milestones in metagenomics. Figure adapted from Escobar-Zepeda et al. (2015)

various applications in biotechnology, healthcare, agriculture, ecology and environment.

13.2 Bioinformatics Approaches for Metagenomics

Two major approaches are used for metagenomics. The first one is called marker-gene metagenomics, and the second approach is referred to as the shotgun metagenomics. The first approach relies on the use of marker genes and an amplicon sequencing method to reassemble the taxonomic structure of a microbial community (Oulas et al. 2015). The marker genes include the conserved or house-keeping genes which are found in a wide range of organisms such as 16S rRNA (Dudhagara et al. 2015a, b, c; Ghelani et al. 2015), 18S rRNA and ITS (Dudhagara et al. 2015a, b, c). The 16S rRNA gene fits very well in this category due to its ubiquitous presence in prokaryotes, being an integral part of the ribosome machinery. The 16S rRNA gene also serves as a molecular clock due to its constant rate of evolution and consists of a unique combination of hypervariable regions as well as conserved regions. The conserved regions allow using universal primers for amplification, while the hypervariable regions provide the variation required to differentiate the microorganisms (Janda and Abbott 2007). However, this method requires a well established, robust and time-consuming procedure apart from its limited ability to analyze only 10–100 cloned per sample. This method, though useful for limited sized samples, is not very much applicable to study large communities. Another problem in using the 16S rRNA gene as a marker is that the annotation is based on their putative association with a taxon defined as an operational taxonomic unit (OTU). OTUs are easy to analyze at the phyla or genera level but difficult to define precisely at the species level. Also, specific genes are predicted based on OTUs instead of their direct sequencing. Moreover, events of horizontal gene transfer and the existence of numerous bacterial strains (Poretsky et al. 2014; Konstantinidis and Tiedje 2007; Konstantinidis and Stackebrandt 2013), impairs with direct gene identification, potentially limiting the understanding of a microbiome.

The whole-genome shotgun sequencing (WGS) approach enables direct sequencing of DNA libraries using random primers (Ranjan et al. 2016). Random primers allow sequencing of overlapping regions and ensure adequate coverage to get a fair idea of the entire community's structure (Patel et al. 2015; Mangrola et al. 2015). With the WGS strategy, taxa can be defined more precisely at the species level. WGS approach also enables us to get novel insights into structural and functional aspects of microbial communities (Singh et al. 2009). However, WGS is expensive and requires very extensive data analysis (Kuczynski et al. 2012; Sims et al. 2014; Luo et al. 2013, 2014). As discussed previously, a variety of NGS technologies have been introduced recently to sequence metagenomes from diverse strata.

The result of the NGS experiment is a set of sequence reads which are processed in many ways which have been reviewed earlier (Dudhagara et al. 2015a, b, c;

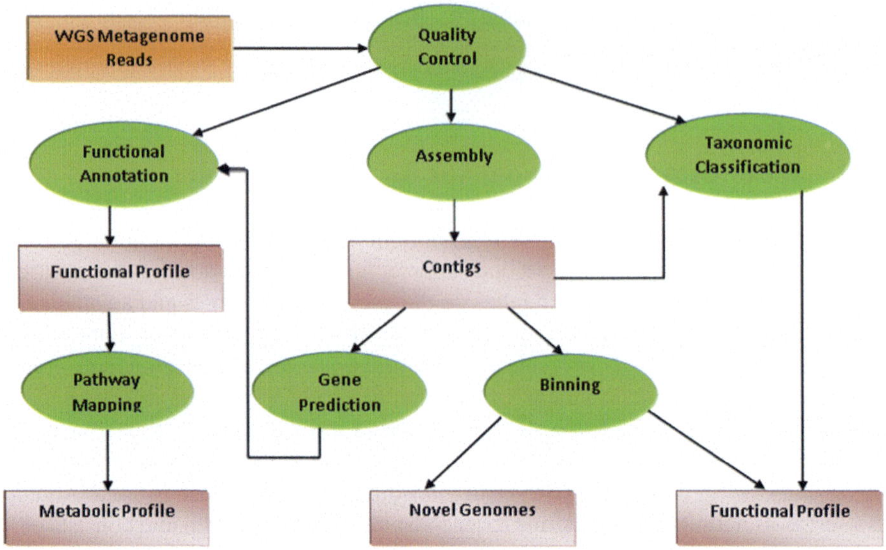

Fig. 13.2 Schematic overview of various steps in metagenomic data analysis workflow. Square boxes represent data and results, and oval boxes represent processing steps. Figure adapted from Jünemann et al. (2017)

Escobar-Zepeda et al. 2015; Jünemann et al. 2017) and can be briefly summarized as in Fig. 13.2. The detailed description of the steps involved is discussed as under.

13.2.1 Quality Control

Assessing and monitoring the quality of reads from any NGS experiment is one of the most crucial steps before proceeding to the analysis of data. Each sequencing technology has its mechanism of detecting the nucleotides from the sample DNA polymer (s) and is therefore prone to errors in detection (Jünemann et al. 2017). The error rate varies with the sequencing platform and negatively affects the identification of microbial communities (Luo et al. 2012a, b). To address the issue of errors, specific file formats have been developed which provide information about the quality, mostly in the form of a Phred score. Phred score provides a quality score for each base, in terms of the estimation of the probability of error (Cock et al. 2010; Table 13.1). The most common of the NGS data file formats include the FASTQ which is an extension of the FASTA format containing the Phred score for each base in the DNA sequence. This format also provides an option to include comments. An overview of the standard file formats is given in Table 13.2. A variety of programs have been available to perform quality analysis of NGS data and provide information including number or read length, GC content, sequence duplication levels, and overrepresented sequences. Some of them are more sophisticated to allow

Table 13.1 Summary of Phred quality score and corresponding base call accuracy

Phred quality score	Probability of incorrect base call	Base call accuracy
10	1 in 10	90%
20	1 in 100	99%
30	1 in 1000	99.9%
40	1 in 10,000	99.99%
50	1 in 100,000	99.999%

Q30 refers to 99.9% accuracy. An error rate lesser than 0.1% is generally considered to be a benchmark for quality (http://www.illumina.com/documents/products/technotes/technote_Q-Scores.pdf). Information adapted from Endrullat et al. (2016)

Table 13.2 An overview of various NGS file formats

Name of file format	Comments	Reference
FASTQ	Stores sequence information, per-base quality and comments (optional)	Cock et al. (2010)
QUAL	Stores sequence information and per-base quality	–
Variant call format (VCF)	Stores information about Indels (insertions/deletions), single nucleotide polymorphisms (SNPs) along with detailed annotation	Danecek et al. (2011)
Sequence alignment/map (SAM) format	Stores read alignments against the reference sequence	Li et al. (2009)
Binary alignment/map (BAM) format	Binary (compressed) version of the SAM format	Li et al. (2009)

modification of sequence reads including trimming, removal of adapters and quality filtering. (Escobar-Zepeda et al. 2015).

13.2.1.1 FASTQC

It is one of the most common and essential tools for evaluating the quality of sequencing data using statistical tests and provides a quick impression of the quality of the data and also indicates the areas of the data having some problems (http://www.bioinformatics.babraham.ac.uk/projects/fastqc/). FASTQC supports SAM, BAM and FASTQ formats and generates summary graphs, tables which can be exported to HTML.

13.2.1.2 NGS QC Toolkit

This toolkit is a user-friendly, standalone set of command line tools for quality control as well as analysis of the sequence reads generated using Illumina and Roche 454 platforms (Patel and Jain 2012). Like FASTQC, this toolkit also generates output in the form of tables and graphs and helps in the filtering of high-quality sequence data. Additionally, it also offers a set of tools which can be used to perform a basic analysis of NGS data.

13.2.1.3 Meta-QC-Chain

This is another dedicated platform for quality control and analysis of NGS data (Zhou et al. 2014). It can be used to retrieve high-quality reads, identify and quantify the source of contaminations and filter contaminating reads. This server (http://www.computationalbioenergy.org/qc-chain.html) also performs mapping of the reads against 18S rRNA databases to detect and remove contaminant sequences of eukaryotic origin.

13.2.1.4 Genome Analysis Toolkit (GATK)

The GATK is particularly useful when dealing with the analysis of variant Calling Format (VCF) data and has become an industry-standard in studies including SNPs, indels variant calling, copy number variations, structural variations, among others. (McKenna et al. 2010). The GATK (https://gatk.broadinstitute.org/hc/en-us) also provides many utilities to carry out quality control tasks.

13.2.2 Assembly

After quality control, the next step is the assembly of reads which is nothing less than a jigsaw puzzle due to the following challenges (El-Metwally et al. 2013):

(a) Placing each piece (read) in the correct position.
(b) Very large number of pieces in the puzzle makes it difficult to determine their correct position.
(c) Ambiguity in positioning similar pieces as many pieces share similar locations.

Metagenome assembly poses additional problems due to biological complexity as the metagenome sequence reads often contains a mixture of unambiguous and ambiguous sequences belonging to a variety of genomes differing in sequence coverage (Jünemann et al. 2017). This mixing of genomic elements makes it difficult to differentiate regions of homology of different strains as well as to differentiate between repetitive and paralogous regions thereby increasing the chance of intergenomic chimeric assemblies or intragenomic misassemblies (Luo et al. 2012a, b; Mende et al. 2012).

Genome assembly may be carried out by using one of the two strategies, i.e. the *de novo* and the comparative assembly approach. In *de novo* assembly, we do not have any guidance or map to refer to carry on the assembly process. This approach is suitable to generate strict novel assemblies, particularly when we lack the availability of reference genomes (Martin and Wang 2011). In the comparative approach, we take advantage of already existing reference genomes from the organism of interest or closely related species. These reference genomes act as a guide to correctly assemble the fragments. This approach is also known as reference-dependent assembly and offers a lot of resequencing applications (Pop et al. 2004).

In a nutshell, the assembly starts with a set of short reads. A computer algorithm called an assembler joins the long contigs together after a thorough quality check.

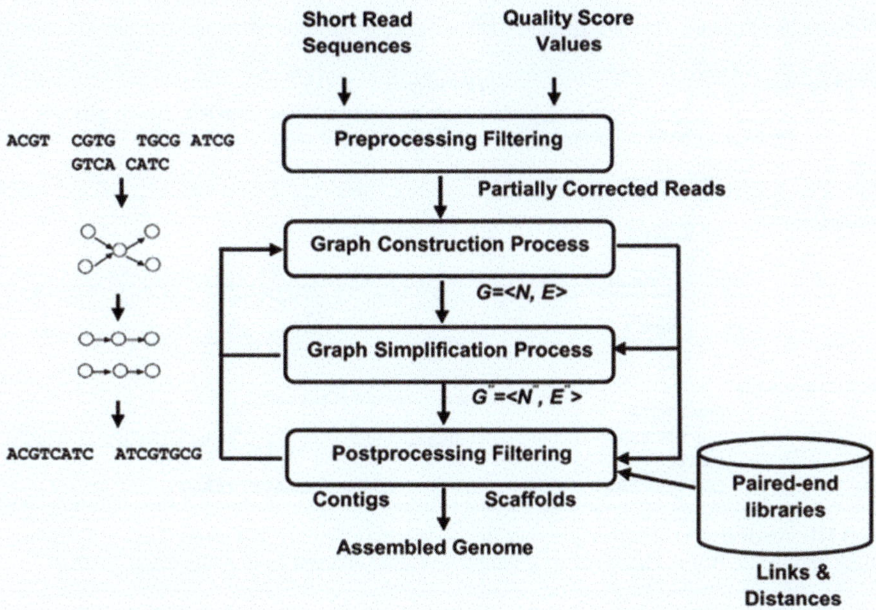

Fig. 13.3 Schematic description of the four stages of NGS Genome Assembly process. Note: G'' is a simplified version of graph G with N nodes and E edges. Figure adapted from El-Metwally et al. (2013)

The contigs thus formed are then aligned to build longer *contigs* called scaffolds (El-Metwally et al. 2013). The entire process is outlined in Fig. 13.3.

The assemblers for single genome or metagenome sequences can be broadly classified into two types, Overlap Layout Consensus (OLC) and de Bruijn Graph (dBg) assemblers, based on the technology used. In OLC, overlaps are identified between reads which are then used to construct an overlap graph (Fig. 13.4a). Based on overlaps (shown as dashed lines), reads are laid-out into contigs. Finally, the most likely sequence is chosen to construct a consensus sequence. In dBg assembly, firstly, reads are divided into k-mers (length k-substrings of the reads) by choosing a sliding window of size *k* across the reads (Fig. 13.4b). These k-mers represents vertices of the dBg with edges connecting overlapping k-mers. Polymorphisms (shown in red colour) form branches in the graph. A record is kept of how many times a k-mer is seen (shown as numbers above k-mers). The contigs are built by walking the graph from edge nodes (Ayling et al. 2020). A detailed description of assemblers based on the above two strategies is beyond the scope of this chapter; however, Table 13.3 contains the list of various assemblers for reference.

Assembly must be evaluated for quality before going for downstream analysis. Several dedicated software tools are available for this task, for example, MetaQUAST (Mikheenko et al. 2016), which is a modified version of the genome assembly evaluation tool QUAST (Gurevich et al. 2013). MetaQUAST not only provides common quality metrics but also performs a reference-based evaluation to

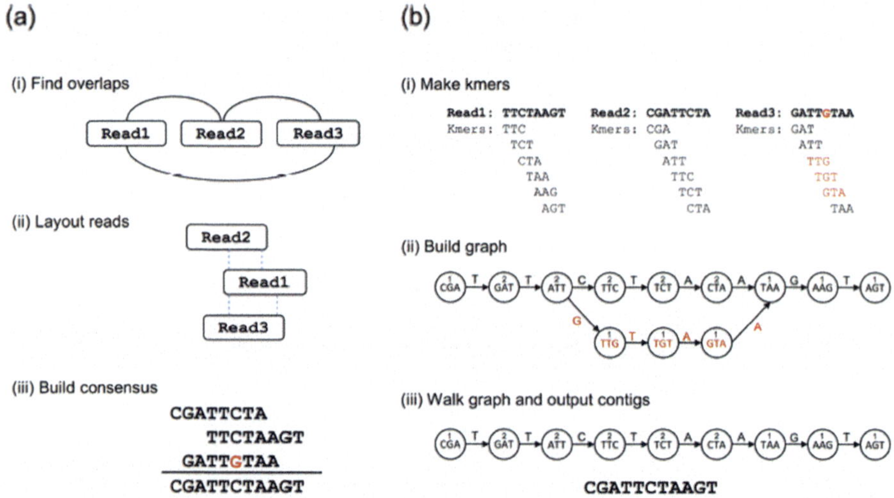

Fig. 13.4 An overview of two major strategies of genome assembly (**a**) Overlap Layout Consensus (OLC) and (**b**) de Bruijn Graph (dBg) assemblers. Figure adapted from Ayling et al. (2020)

identify misassemblies. This is achieved by mapping contigs back to the reference genomes and reporting quality statistics separately for each reference genome.

13.2.3 Binning

Genome binning is one of the initial steps in downstream processing and analysis of metagenome assemblies which contain sequence fragments of variable length having their origin in distinct genomes. In genome binning, the related fragments are clustered by utilizing different techniques. These clusters are called bins. If the reference genomes are available, the sequences are aligned against their reference genomes to accomplish genome binning. However, in the absence of reference genomes, inherent sequence composition measures such as GC content or k-mer frequencies are used to achieve binning, as these properties vary across organisms (Jünemann et al. 2017). These composition-based measures often make use of machine learning approaches. The k-mer frequency methods rely on the assumption that k-mers frequency distributions of whole genomes or their fragments are unique to each genome. Also, it has been observed that the relative k-mers show intragenome similarities and intergenome differences (Qian and Comin 2019). In this strategy, sequences are represented as vectors using short words (k-mers). These k-mers acts as genomic signatures and are then used to measure similarity among different *contigs*. The software PhylophytiaS (McHardy et al. 2007), TETRA (Teeling et al. 2004), MetaCon (Qian and Comin 2019) and MetaclusterTA (Wang et al. 2014) are based on classification by the composition of k-mers. Amphora2

Table 13.3 Examples of software used at various stages of metagenomic data analysis and metaprofiling

Software name	Links	Applications/features	References
BLASTX	https://blast.ncbi.nlm.nih.gov/Blast.cgi?LINK_LOC=blasthome&PAGE_TYPE=BlastSearch&PROGRAM=blastx	Homology search by using translated nucleotide sequences (six-frame) against a database of amino acid sequences	Altschul et al. (1997)
TETRA	http://omictools.com/tetra-s1030.htm	Taxonomic classification by utilizing tetranucleotide patterns	Teeling et al. (2004)
CD-HIT	http://weizhongli-lab.org/cd-hit/	DNA/RNA/Protein sequences are clustered and compared	Li and Godzik (2006)
PhylophytiaS	http://omictools.com/phylopythia-s1455.html	Utilizes reference-genome signature for composition-based classification of sequences	McHardy et al. (2007)
CARMA	http://omictools.com/carma-s1021.html	Uses conserved domains (Pfam) for phylogenetic classification of sequence reads	Krause et al. (2008)
MetaORFA	NA	Uses predicted ORFs for the assembly of peptides	Ye and Tang (2009)
MG-RAST	http://metagenomics.anl.gov/	An open web resource for comparative metagenomics, taxonomic and functional annotation. Also, provides a graphical interface and a web portal	Meyer et al. (2008)
MinPath	http://omics.informatics.indian.edu/MinPath/	Uses protein family prediction data for reconstructing pathways	Ye and Doak (2009)
ShotgunFunctionalizeR	http://shotgun.math.chalmers.se/	Metagenomic functional comparison of individual gene sequences (COG and EC numbers). Also provides analysis of complete pathways	Kristiansson et al. (2009)
DiScRIBinATE	http://metagenomics.atc.tcs.com/binning/DiScRIBinATE/	Utilizes the best hits of BLASTx for taxonomic classification	Ghosh et al. (2010)

(continued)

Table 13.3 (continued)

Software name	Links	Applications/features	References
MetaGeneMark	http://exon.gatech.edu/index.html	Utilizes the heuristic model for prediction of coding sequences from metagenomic reads	Zhu et al. (2010)
FragGeneScan	http://sourceforge.net/projects/fraggenescan/	Predicts coding sequences from shorter reads	Rho et al. (2010)
Galaxy portal	https://usegalaxy.org/	Provides a graphical interface and a free web portal for various computational tools	Afgan et al. (2018)
METAREP	http://www.jcvi.org/metarep	Analysis and comparison of metagenomics datasets	Goll et al. (2010)
HMMER3	http://hmmer.janelia.org/	Performs sequence alignment using the hidden Markov models	Eddy (2011)
MetaPath	http://metapath.cbcb.umd.edu/	Analyzes relative abundance of metabolic pathways in environmental samples	Liu and Pop (2011)
ProViDE	http://metagenomics.atc.tcs.com/binning/ProViDE	Analysis of viral diversity in metagenomic samples	Ghosh et al. (2010)
SPAdes	http://bioinf.spbau.ru/spades	Single-cell assembly and analysis	Bankevich et al. (2012)
NGS QC	http://www.nipgr.res.in/ngsqctoolkit.html	Performs QC analysis in parallel environments	Patel and Jain (2012)
MOCAT	http://vm-lux.embl.de/~kultima/MOCAT2/index.html	Quality control, taxonomic annotation, classification of marker genes, prediction of coding regions	Kultima et al. (2012)
Amphora and Amphora2	http://pitgroup.org/amphoranet/	Metagenomic phylotyping by single-copy phylogenetic marker genes classification	Wu and Eisen (2008), Wu and Scott (2012)

Bowtie	http://bowtie-bio.sourceforge.net/index.shtml	Uses the burrows-wheeler transform for quick alignment of shorter reads with references sequences	Langmead and Salzberg (2012)
IDBA-UD	http://i.cs.hku.hk/~alse/hkubrg/projects/idba_ud/	Performs *de novo* assembly of metagenomic sequences with uneven depth	Peng et al. (2012)
MetaVelvet	http://metavelvet.dna.bio.keio.ac.jp/	Performs de-novo assembly of shorter metagenomic sequence reads	Namiki et al. (2012)
RayMeta	http://denovoassembler.sourceforge.net/	Assembler of *de novo* of metagenomic reads and taxonomy profiler by ray communities	Boisvert et al. (2012)
GlimmerMG	http://www.cbcb.umd.edu/software/glimmer-mg/	Uses unsupervised clustering algorithm for prediction of coding sequences from the metagenomic sequence data	Kelley et al. (2012)
IMG/M	https://img.jgi.doe.gov/cgi-bin/m/main.cgi	An open web resource for gene distribution, comparative metagenomics, taxonomic and functional annotation. Also, provides a graphical interface and a web portal	Markowitz et al. (2012)
METAGENassist	http://www.metagenassist.ca	Comprehensive web server for comparative metagenomics which can also perform an automated taxonomy-to-phenotype mapping	Arndt et al. (2012)
MEGAN	http://ab.inf.uni-tuebingen.de/software/megan5/	Uses the BLAST result for comparative metagenomics. Also performs the taxonomic and functional analysis of metagenomic reads; graphical interface	Huson and Weber (2013)
PICRUSt	http://picrust.github.io/picrust/	Uses the information from 16S rRNA metaprofiling projects for prediction of metabolic potential	Langille et al. (2013)

(continued)

Table 13.3 (continued)

Software name	Links	Applications/features	References
Genometa	http://genomics1.mh-hannover.de/genometa/	Predicts classification and performs functional annotation of short-reads metagenomic data. Also provides a graphical interface	Davenport and Tümmler (2013)
MetagenomeSeq	http://bioconductor.org/packages/release/bioc/html/metagenomeSeq.html	Provides a bioconductor package for evaluation of relative abundance of 16S rRNA gene in meta-profiling data	Paulson et al. (2013)
Parallel-meta	http://www.computationalbioeergy.org/parallel-meta.html	Uses the best BLAST hits for gene annotation. Also used metagenome reads for annotation of rRNA genes	Su et al. (2014)
MetaclusterTA	http://i.cs.hku.hk/~alse/MetaCluster	Performs the binning of reads and sequence contigs for taxonomic annotation.	Wang et al. (2014)
MaxBin	http://sourceforge.net/projects/maxbin/	Performs binning of short reads and contiguous sequences using unsupervised binning algorithms	Wu et al. (2016)
Phyloseq	https://joey711.github.io/phyloseq/	Uses an R, bioconductor package. Performs processing and analysis of diversity. Also, provides graphics output	McMurdie and Holmes (2015)
GroopM	http://ecogenomics.github.io/GoopM/	Uses contiguous sequence coverages for identification of population genomes	Imelfort et al. (2014)
OneCodex	https://www.onecodex.com	A sensitive and accurate data platform for genomic microbial identification	Minot et al. (2015)
MetaPhlAn2	http://segatalab.cibio.unitn.it/tools/metaphlan2/	Enhanced metagenomic taxonomic profiling. Identification of specific strains, tracking strains across samples for all species	Truong et al. (2015)

CLARK	http://clark.cs.ucr.edu/	Provides a fast, versatile and accurate sequence classification method,. Very useful for applications in metagenomics and genomics	Ounit et al. (2015)
Tax4Fun	http://tax4fun.gobics.de/	Uses 16S sequence data via a free R package for prediction of functional abilities microbial communities	Aßhauer et al. (2015)
DIAMOND	https://github.com/bbuchfink/diamond	A very fast method that uses spaced seeds with a reduced amino acid alphabet	Buchfink et al. (2015)
CheckM	https://kbase.us/applist/apps/kb_Msuite/run_checkM_lineage_wf/release?gclid=CjwKCAjw1cX0BRBmEiwAy9tKHu5KgOt2h63dG_lQr2fRCvu8Z1IDoyVomAonliQ-34ckMGYj5sen9BoC8DAQAvD_BwE	Uses a set of specific marker genes for assessing the quality of a genome and information about the proximity of these genes	Parks et al. (2015)
FastQC	http://www.bioinformatics.babraham.ac.uk/projects/fastqc	Uses modular options for performing the quality control of high-throughput sequencing data. Gives graphic results of quality per base sequence, GC-content, N numbers, duplication, and over represent	Andrews (2015)
MEGAHIT	https://hku-bal.github.io/megabox	Performs assembly tasks on metagenomic sequence datasets of hundreds of Giga base-pairs long. Provides time-and memory-efficient performance on a single server	Li et al. (2016)
MEGAN-CE	https://github.com/danielhusonmegan-ce	An open-source program which performs interactive exploration and analysis of large-scale microbiome sequencing data	Huson et al. (2016)
Piphillin	http://piphillin.secondgenome.com/	A software package that predicts functional metagenomic content based on the frequency of detected 16S rRNA gene sequences corresponding to genomes in regularly updated, functionally annotated genome databases	Iwai et al. (2016)

(continued)

Table 13.3 (continued)

Software name	Links	Applications/features	References
DUDes	http://sf.net/p/dudes	Performs reference-based taxonomic profiling that introduces a novel top-down approach to analyze metagenomic high-throughput sequencing samples	Piro et al. (2016)
MetaQuast	http://bioinf.spbau.ru/metaquast	A modification of QUAST, that uses contig-alignment data for genome assembly, using a reference	Mikheenko, et al. (2016)
Kallisto	https://github.com/pachterlab/kallisto	It quantifies abundances of transcripts from RNA-Seq data and can be used for taxonomy profiling	Bray et al. (2016)
MetaBAT	https://bitbucket.org/berkeleylab/metabat	Metagenome binning with abundance and tetra-nucleotide frequencies	Kang et al. (2015)
MetaSort	https://sourceforge.net/projects/metasort/	Provides a sorted mini-metagenomic approach based on single-cell sequencing and flow cytometry methodologies. Also performs efficient recovery of high-quality genomes from the binned mini metagenome using complementarity with the original metagenome	Ji et al. (2017)
MSPminer	www.enterome.com/downloads	A computationally efficient software tool that reconstitutes metagenomic species Pan-genomes (MSPs) by binning co-abundant genes across metagenomic samples	Plaza Oñate et al. (2019)
MetaBAT2	https://bitbucket.org/berkeleylab/metabat	Uses a new adaptive binning algorithm to eliminate manual parameter tuning	Kang et al. (2019)
Kraken2	https://ccb.jhu.edu/software/kraken2/index.shtml	Provides a fast taxonomic classification of metagenomic sequences	Wood et al. (2019)

CosmosID	https://www.cosmosid.com/	Provides fast, reliable bacterial detection and identification from metagenomic shotgun sequencing data derived from sonicate fluid for the diagnosis of PJI	Yan et al. (2019)
GraphBin	https://github.com/Vini2/GraphBin	A new binning method that makes use of the assembly graph and applies a label propagation algorithm to refine the binning result of existing tools	Mallawaarachchi et al. (2020)
DeepMAsED	https://github.com/leylabmpi/DeepMAsED	a deep learning approach for identifying misassembled contigs without the need for reference genomes	Mineeva et al. (2020)
Pavian	http://github.com/fbreitwieser/pavian	a web application for exploring classification results from metagenomics experiments	Breitwieser and Salzberg (2020)
rnaSPAdes	http://cab.spbu.ru/software/rnaspades/	SPAdes-based *de novo* transcriptome assembler. It can support hybrid assembly based on short and long reads. It is distributed as a part of SPAdes assembler	Bushmanova et al. (2019)

(Wu and Scott 2012) and MaxBin (Wu et al. 2016) and use more than one measures of sequence composition such as GC composition, k-mer signatures, single-copy marker genes and sequence coverage information. For example, MaxBin uses tetranucleotide frequency as well as contig abundances to complete the binning process more precisely.

In an alternate strategy, based on the alignment of reads to reference genomes, several software tools have also been developed including those based on Burrows-Wheeler Transform indexes such Burrows-Wheeler Aligner (BWA) and Bowtie (Langmead and Salzberg 2012; Li and Durbin 2010). These methods can quickly and accurately estimate the taxonomic abundance via species-richness by mapping sequence reads directly to unique reference genomes or many pan genomes (concatenated genomes) or sequences (Escobar-Zepeda et al. 2015).

13.2.4 Gene Prediction and Functional Annotation

After the genome binning step, data needs to be analyzed for taxonomic classification as well as characterization. In one of the approaches called fragment recruitment, the genomic assemblies are mapped to the databases of annotated genes or proteins. Matches with good coverage are then subjected to more sophisticated analysis, including mapping to metabolic pathways. In another strategy called *de novo* annotation, metagenomic reads of metagenome assemblies (contigs, scaffolds) are annotated from scratch starting from gene prediction. Many gene prediction pipelines are available for prediction of genes belonging to a single genome. However, these are not well suited to handle the diversity of metagenomic datasets in terms of composition, length and the error-rates in sequences (Hoff 2009). In recent years, many *de novo* gene prediction programs have been developed to suit annotation of metagenomic data sets including—MetaGene (Noguchi et al. 2006), Orphelia (Hoff et al. 2009), MetaGeneMark (Zhu et al. 2010), Prodigal in metagenomic mode (Hyatt et al. 2012), MetaGun (Liu et al. 2013), among others. These predicted coding sequences (CDS) can be further analyzed using homology-based mapping (using BLAST) against specialized databases such as EggNOG (Huerta-Cepas et al. 2016), COG (Tatusov et al. 2000) or KEGG orthology groups (Kanehisa et al. 2016a, b). Also, Hidden Markov Model (HMM) based methods such as HMMER (Eddy 2011), PFAM (El-Gebali et al. 2019) can be very useful assigning protein sequences to their respective protein families. These software tools have been described in the following sections, while a list of additional platforms is provided in Table 13.3.

13.2.4.1 MetaGene

MetaGene is a prokaryotic gene finding program. It takes anonymous genomic sequences as input and processes them through a two-stage process. In the first stage, it extracts all possible ORFs and scores them using length and base composition. In the second stage, orientation scores and distances of neighboring ORFs are combined with scores of individual ORFs to find the optimal (high-scoring) combination of ORFs. The scoring is done using Log-odds ratios obtained by dividing the frequency of occurrence of an event in protein-coding ORFs by its frequency in random ORFs. The statistical parameters used include ORF length distributions, di-codon frequencies, orientation-dependent distances and frequencies of orientations of neighboring ORFs and distance distributions from an annotated start codon to the leftmost start codon. MetaGene (http://metagene.nig.ac.jp/metagene/metagene.html) is freely available for academic users.

13.2.4.2 Orphelia

Orphelia is a gene-prediction tool for short metagenomic DNA sequences from environmental samples with an unknown phylogenetic origin. Orphelia utilizes a two-stage process based on machine learning. The first stage begins with the prediction of ORFs, which are then subjected to feature extraction using linear discriminants. Subsequently, these features are combined using an artificial neural network to predict genes. Orphelia accurately predicts genes even from species that were not a part of the training set. Orphelia can be used as a web server (http://orphelia.gobics.de/) and can also be used as a standalone application on Linux operating systems.

13.2.4.3 MetaGeneMark

MetaGeneMark is an *ab initio* gene prediction method for anonymous short DNA sequences. It uses a heuristic approach to estimate parameters from linear dependencies between oligonucleotides frequencies from protein-coding regions and genome nucleotide compositions (e.g. GC content). MetaGeneMark (http://exon.gatech.edu/meta_gmhmmp.cgi) is freely accessible in form of a web server.

13.2.4.4 Prodigal in Metagenomic Mode

Prodigal *(PROkaryotic Dynamic Programming Genefinding Algorithm)* predicts protein-coding genes from bacterial and archaeal genomes. In normal mode, Prodigal uses unsupervised machine learning strategy and trains itself automatically from the properties of the genome sequence including genetic code, start codon usage frequencies, codon statistics and Ribosomal Binding Site (RBS) motif usage. The normal mode is suitable for finished genomes, draft genomes (reasonable quality), and genomes of big viruses. It has two other modes called anonymous and training modes. The anonymous mode uses pre-calculated training files as input and is suitable for metagenomes, draft genomes (low quality), small viruses, and small plasmids. The training mode works similar to normal mode but saves training files for future use. Prodigal is very fast and accurate. The gene predictions are provided in standard formats such as GFF3, Genbank, Sequin table along with detailed summary statistics. Prodigal is a standalone open-source application (https://github.com/hyattpd/Prodigal).

13.2.4.5 MetaGun

MetaGun (http://cqb.pku.edu.cn/ZhuLab/MetaGUN/index.html) utilizes the Support Vector Machine (SVM) approach to identify genes from metagenomic fragments. In the three-stage SVM methodology, the first stage classifies the input metagenomic fragments into phylogenetic groups. In the second stage, SVM classifiers are used to identify protein-coding sequences independently for each group by integrating entropy density profiles (EDP) of codon usage, translation initiation site (TIS) scores and open reading frame (ORF) lengths. In the third stage, the TISs are adjusted by employing a modified version of MetaTISA (Metagenomic Translation Initiation Site Annotator). MetaGun is freely available to download and use as a standalone application.

13.2.4.6 Clusters of Orthologous Groups (COG)

COG (http://www.ncbi.nlm.nih.gov/COG) is used to compare predicted and known proteins in completely sequenced genomes to infer a set of orthologs. Each COG contains a group of proteins found to be orthologous across at least three lineages and likely corresponds to an ancient conserved domain. Due to its smaller size as compared to the NCBI nr database, it helps in rapid functional annotation of microbial genes.

13.2.4.7 EggNOG

EggNOG is an open-source database of gene evolutionary histories, orthology relationships and functional annotations. The present version, 5.0 (Huerta-Cepas et al. 2016) consists of datasets from 5090 organisms and 2502 viruses amounting. It consists of over 4.4 million orthologous groups (OGs) distributed across 379 taxonomic levels which were computed using sequence alignments, HMM models, phylogenies and functional descriptors. All precomputed data are open to download or can be accessed via API queries at http://eggnog.embl.de.

13.2.4.8 KEGG Orthology (KO) Database

KO (https://www.genome.jp/kegg/ko.html) database is a functional database denoted in terms of functional orthologs. A functional ortholog is manually defined in the context of KEGG molecular networks, namely, BRITE hierarchies, KEGG pathway maps and KEGG modules.

13.2.4.9 HMMER

HMMER (http://hmmer.org/) is a robust method for homology search and construction sequence alignments using probabilistic models celled profile hidden Markov models (profile HMMs). Its profile based probabilistic nature makes it very sensitive to detect remote homologs. In addition to profile based search, it also offers sequence-based search. HMMER is a bundle of small programs which are used to perform various tasks including sequence search, alignment building and profile generation. HMMER can be downloaded and used as a standalone application. HMMER is a part of the PFAM database. It is also used with many databases of Interpro.

13.2.4.10 PFAM

PFAM (https://pfam.xfam.org/) is a database of protein families represented in the form of multiple sequence alignments and hidden Markov models. In Pfam, the related protein families are further clubbed to generate higher-level groups called *clans*. The families in a clan are included based on the sequence similarity, structure or profile-HMMs. PFAM provides various search options, including keyword search and sequence search. PFAM helps in functional annotation of proteins based on the concept of conserved domains.

In addition to the tools mentioned above for gene prediction and functional annotation, there are some integrated platforms capable of taking care of various aspects of metagenome data processing, assembly, annotation, among others. A few of these have been described below.

13.2.4.11 MG-RAST

Metagenomics Rapid Annotation using Subsystem Technology (MG-RAST) is one of the earlier, open-source, user-friendly platforms introduced for the analysis of metagenomic data (Meyer et al. 2008). The latest version of MG-RAST (4.0.3) harbors 4, 20,411 metagenomes containing 1639 billion sequences and consists of 31,406 registered users (as on 18-04-2020). The raw metagenomic data can be submitted in standard sequence formats including in FASTA, AFF and FASTQ formats along with detailed information about the sample. Data and metadata uploading can be done using the web server (https://www.mg-rast.org/) or by using the command-line version. The data submitted is treated as private unless the user chooses to make it public. MG-RAST employs a multi-step workflow including protein prediction, clustering, quality control and similarity-based annotation on nucleic acid sequence datasets using many bioinformatics tools (Meyer et al. 2008). Automated annotation is performed by utilizing an extensive collection of reference datasets. The web interface can be used to perform various other tasks such as phylogenetic analysis, functional annotation, metabolic pathway profiling apart from comparing two or more metagenomes.

13.2.4.12 IMG/M

The Integrated Microbial Genomes and Microbiomes system (IMG/M: https://img.jgi.doe.gov/m/) is a data management, storage and analysis system for metagenomes curated by the Joint Genome Institute (JGI) of the United States Department of Energy (DOE). Its latest version 5.0 is populated with annotated datasets, organized into various categories including archaea, bacteria, eukaryotes, plasmids, viruses, genome fragments, metagenomes, cell enrichments, single-particle sorts, and metatranscriptomes (Chen et al. 2019). The available datasets consist of the internal datasets generated by the JGI (DOE) as well the datasets uploaded by external scientists. In addition, data from public databases (e.g. NCBI) are also included in the source datasets. The IMG annotation pipeline is used to process all submissions before loading them into the IMG data warehouse. The IMG web server offers a variety of analytical and visualization tools to carry out the comparative analysis of single genomes as well as metagenomes. IMG/M provides free access to all publicly available genomes available in its data warehouse. However, registered users are also allowed to access individual genomes through the dedicated expert review (ER) system (IMG/MER: https://img.jgi.doe.gov/mer/). Registered users can further store their private datasets in the workspace for sharing and further analysis (Chen et al. 2019).

13.2.4.13 Galaxy

Galaxy (https://usegalaxy.org) is an open source web-based platform for scientific analysis of genomic data. This platform is being utilized by a very large number of researchers across the globe for the analysis of large biomedical datasets containing data related to genomics, metagenomics, proteomics, metabolomics and imaging experiments. The Galaxy project was initiated in 2005 to address three critical challenges of data-driven biomedical science:

(a) Ensuring analyses are entirely *reproducible*
(b) Making analyses *accessible* to all researchers
(c) Making it simple to *communicate* analyses so that they can be reused and extended (Afgan et al. 2018).

The Galaxy ToolShed provides more than 5500 tools and pipelines for analyzing datasets. This collection of programs includes those for general text manipulation, genomic file manipulation, genomics analysis, metagenomic analysis, toolkits, statistics and visualization. Metagenomic analysis programs offer a variety of options to carry out different tasks related to read quality checking, read assembly, binning, functional annotation, among others.

Table adapted from Escobar-Zepeda et al. (2015).

13.3 Challenges and Opportunities in Metagenomics

Metagenomics has opened many new avenues for studies related to microbial communities. We have reached a level, where we are looking forward to combining metagenomics data with other complex datasets such as biodiversity data (e.g. from 16S rRNA gene amplicon sequencing), *in situ* expression data (metatranscriptomics and metaproteomics) and environmental factors to take a holistic view of the ecosystem (Simon and Daniel 2010; Shi et al. 2011; Teeling et al. 2012). Attempts have also been made to integrate data from metagenomic studies with the metabolome data (Turnbaugh and Gordon 2008). However, such complex applications often bring a plethora of challenges related to experimental design, data processing and bioinformatics analysis. We are still short of standard methods or universal tools to answer all questions related to metagenomics (Escobar-Zepeda et al. 2015). In fact, due to the lack of standards, reproducibility is challenging to achieve. It is important to note that every metagenome project has its specific requirements, so utmost care should be taken to choose the experimental design, sequencing technology and computational tools for downstream analysis. A metagenome usually represents a snapshot of a community at a certain point of time when its DNA is obtained. Therefore, a robust and detailed experimental design is required to catch the complete population dynamics over a period of time to make real observations. Such an experimental approach needs to integrate different approaches such as culture methods, DNA and RNA analysis, protein studies, and if possible, the metabolic profiles (Escobar-Zepeda et al. 2015). So, standardization of methods, efficiently integrating various approaches is an excellent opportunity to look upon. Furthermore, fine-tuning the methodology to get a real picture of the natural environment of microbial communities by studying their dynamics as well as interactions with other stakeholders is going to be an important research area.

Challenges are also faced in the sequencing and assembly of samples. Modern-day sequencers have brought down the cost of sequencing to much affordable level and can generate a vast amount of data. However, the cost factor still limits the scale of our experimental design, particularly for complex microbial communities due to huge sequencing requirements, which enormously increases the overall cost. Additionally, it is often difficult to have sufficient representation from a large number of organisms available in many ecosystems. This variable relative abundance of different community members within a population causes variability in the representation of genomes. In such a scenario, some genomes may be covered very excessively, while others are only covered by a handful of sequencing reads or none at all. So, more advancement in sequencing technologies is the need of the hour to make them cost-effective, accurate and more sensitive to deal with population bias.

Furthermore, assembling the genomes poses different challenges, especially when the communities are very diverse with none of the members represented significantly. As the assembly of sequence data requires overlaps among reads, less dominant members of a community may require additional sequencing (Howe and Chain 2015). Another challenge for metagenomic assembly is that despite having very robust assembly algorithms and powerful computing resources, assembly of such abundant, complex data can often overwhelm any given computer's memory constraints. This issue arises due to the natural diversity of the community and the variants found within the population. Sequencing errors further exacerbate this issue (Howe and Chain 2015). Bioinformatics algorithms have played a central role in the development of metagenomics, but, keeping in view the above assembly challenges, we have much room to suggest new strategies to create accurate sequence assemblies, taking care of the inherent variability as well as sequencing errors.

The ultimate challenge is the analysis of this enormous data, particularly the lack of appropriate statistical framework. It is unsuitable for applying classical statistics due to enormous dimensionality of data having millions of variables with uneven representation. This data is even not suitable to apply parametric tests after log-transformations as it shows long-tailed distributions. Furthermore, strong interdependence among genes and sometimes species makes a variable selection, a tedious job (Prifti and Zucker 2015). One way to resolve this issue is by reducing the dimensionality of the data. This has been achieved by applying clustering techniques which cluster groups of genes together corresponding to their core genomes based on their abundance profiles or gene assembly (Nagarajan and Pop 2013). Certain events, such as horizontal gene transfer (e.g. transfer of some portion of DNA from a virus to a bacterium), poses another challenge to analysis of metagenomics data. So, we need to develop new techniques to reduce the dimensionality of the metagenomic data and robust algorithms to detect genetic transfer events, including both vertical as well as horizontal.

13.4 Conclusion

Metagenomics is a very dynamic and robust strategy to study the microbial world and has provided significant insights since its inception. We have even been able to study microbial populations which were considered unculturable in the recent past. Advancements in molecular techniques, computational resources and data analytics have made it possible to scale up metagenomics to deal with mega projects which were never thinkable a few decades ago. Recently, we have been able to develop more sophisticated software tools with enhanced performance concerning the complexities of metagenomic data. However, these software tools need further improvement to capture more variations in the data. In recent times, the concept of developing data analysis pipelines by integrating a variety of software tools into a single platform is gaining more importance. Like many other technologies, metagenomics is still developing and therefore faces many challenges. However, one cannot rule out the opportunities this technology offers to study the microbial world in particular, and the environment as a whole. Metagenomics, thus provides a window into a world of unseen microbial diversity which can be explored using biotechnological tools, thereby paving the way to novel scientific, environmental, pharmaceutical and industrial applications.

References

Afgan E, Baker D, Batut B et al (2018) The galaxy platform for accessible, reproducible and collaborative biomedical analyses: 2018 update. Nucleic Acids Res 46:W537–W544

Altschul SF, Madden TL, Schäffer AA et al (1997) Gapped BLAST and PSI-BLAST: a new generation of protein database search programs. Nucleic Acids Res 25:3389–3402

Andrews SC (2015) FastQC v0.11.3. Babraham Bioinformatics, Cambridge, MA

Arndt D, Xia J, Liu Y et al (2012) METAGENassist: a comprehensive web server for comparative metagenomics. Nucleic Acids Res 40:W88–W95

Aßhauer KP, Wemheuer B, Daniel R, Meinicke P (2015) Tax4Fun: predicting functional profiles from metagenomic 16S rRNA data. Bioinformatics 31:2882–2884

Ayling M, Clark MD, Leggett RM (2020) New approaches for metagenome assembly with short reads. Brief Bioinform 21:584–594

Bankevich A, Nurk S, Antipov D et al (2012) SPAdes: a new genome assembly algorithm and its applications to single-cell sequencing. J Comput Biol 19:455–477

Biddle JF, Fitz-Gibbon S, Schuster SC et al (2008) Metagenomic signatures of the Peru margin subseafloor biosphere show a genetically distinct environment. Proc Natl Acad Sci 105:10583–10588

Boisvert S, Raymond F, Godzaridis É et al (2012) Ray Meta: scalable de novo metagenome assembly and profiling. Genome Biol 13:R122

Bray NL, Pimentel H, Melsted P, Pachter L (2016) Near-optimal probabilistic RNA-seq quantification. Nat Biotechnol 34:525–527

Breitwieser FP, Salzberg SL (2020) Pavian: interactive analysis of metagenomics data for microbiome studies and pathogen identification. Bioinformatics 36:1303–1304

Buchfink B, Xie C, Huson DH (2015) Fast and sensitive protein alignment using DIAMOND. Nat Methods 12:59

Bushmanova E, Antipov D, Lapidus A, Prjibelski AD (2019) rnaSPAdes: a de novo transcriptome assembler and its application to RNA-Seq data. Gigascience 8:giz100

Chen I-MA, Chu K, Palaniappan K et al (2019) IMG/M v.5.0: an integrated data management and comparative analysis system for microbial genomes and microbiomes. Nucleic Acids Res 47: D666–D677

Chiu CY, Miller SA (2019) Clinical metagenomics. Nat Rev Genet 20:341–355

Cock PJA, Fields CJ, Goto N et al (2010) The sanger FASTQ file format for sequences with quality scores, and the Solexa/Illumina FASTQ variants. Nucleic Acids Res 38:1767–1771

Danecek P, Auton A, Abecasis G et al (2011) The variant call format and VCFtools. Bioinformatics 27:2156–2158

Daniel R (2005) The metagenomics of soil. Nat Rev Microbiol 3:470–478

Davenport CF, Tümmler B (2013) Advances in computational analysis of metagenome sequences. Environ Microbiol 15:1–5

DeLong EF, Preston CM, Mincer T et al (2006) Community genomics among stratified microbial assemblages in the ocean's interior. Science 311:496–503

Dudhagara P, Bhavsar S, Bhagat C et al (2015a) Web resources for metagenomics studies. Genomics Proteomics Bioinformatics 13:296–303

Dudhagara P, Ghelani A, Bhavsar S, Bhatt S (2015b) Metagenomic data of fungal internal transcribed spacer and 18S rRNA gene sequences from Lonar lake sediment, India. Data Br 4:266–268

Dudhagara P, Ghelani A, Patel R et al (2015c) Bacterial tag encoded FLX titanium amplicon pyrosequencing (bTEFAP) based assessment of prokaryotic diversity in metagenome of Lonar soda lake, India. Genom Data 4:8–11

Eddy SR (2011) Accelerated profile HMM searches. PLoS Comput Biol 7:e1002195

El-Gebali S, Mistry J, Bateman A, Eddy SR, Luciani A, Potter SC, Qureshi M, Richardson LJ, Salazar GA, Smart A, Sonnhammer ELL (2019) The Pfam protein families database in 2019. Nucleic Acids Res 47(D1):D427–D432

El-Metwally S, Hamza T, Zakaria M, Helmy M (2013) Next-generation sequence assembly: four stages of data processing and computational challenges. PLoS Comput Biol 9:e1003345

Endrullat C, Glökler J, Franke P, Frohme M (2016) Standardization and quality management in next-generation sequencing. Appl Transl Genom 10:2–9

Escobar-Zepeda A, Vera-Ponce de Leon A, Sanchez-Flores A (2015) The road to metagenomics: from microbiology to DNA sequencing technologies and bioinformatics. Front Genet 6:348

Ferrer M, Beloqui A, Timmis KN, Golyshin PN (2009) Metagenomics for mining new genetic resources of microbial communities. J Mol Microbiol Biotechnol 16:109–123

Ghelani A, Patel R, Mangrola A, Dudhagara P (2015) Cultivation-independent comprehensive survey of bacterial diversity in Tulsi Shyam Hot Springs, India. Genom Data 4:54–56

Ghosh TS, Haque M, Mande SS (2010) DiScRIBinATE: a rapid method for accurate taxonomic classification of metagenomic sequences. BMC Bioinform 11(7):S14

Gillespie DE, Brady SF, Bettermann AD et al (2002) Isolation of antibiotics turbomycin A and B from a metagenomic library of soil microbial DNA. Appl Environ Microbiol 68:4301–4306

Goll J, Rusch DB, Tanenbaum DM et al (2010) METAREP: JCVI metagenomics reports—an open source tool for high-performance comparative metagenomics. Bioinformatics 26:2631–2632

Gurevich A, Saveliev V, Vyahhi N, Tesler G (2013) QUAST: quality assessment tool for genome assemblies. Bioinformatics 29:1072–1075

Handelsman J (2004) Metagenomics: application of genomics to uncultured microorganisms. Microbiol Mol Biol Rev 68:669–685

Handelsman J, Rondon MR, Brady SF et al (1998) Molecular biological access to the chemistry of unknown soil microbes: a new frontier for natural products. Chem Biol 5:R245–R249

Hoff KJ (2009) The effect of sequencing errors on metagenomic gene prediction. BMC Genomics 10:520

Hoff KJ, Lingner T, Meinicke P, Tech M (2009) Orphelia: predicting genes in metagenomic sequencing reads. Nucleic Acids Res 37:W101–W105

Howe A, Chain PSG (2015) Challenges and opportunities in understanding microbial communities with metagenome assembly (accompanied by IPython Notebook tutorial). Front Microbiol 6:678

Huerta-Cepas J, Szklarczyk D, Forslund K et al (2016) eggNOG 4.5: a hierarchical orthology framework with improved functional annotations for eukaryotic, prokaryotic and viral sequences. Nucleic Acids Res 44:D286–D293

Huson DH, Beier S, Flade I et al (2016) MEGAN community edition-interactive exploration and analysis of large-scale microbiome sequencing data. PLoS Comput Biol 12:e1004957

Huson DH, Weber N (2013) Microbial community analysis using MEGAN. In: Methods in enzymology. Elsevier, pp 465–485

Hyatt D, LoCascio PF, Hauser LJ, Uberbacher EC (2012) Gene and translation initiation site prediction in metagenomic sequences. Bioinformatics 28:2223–2230

Imelfort M, Parks D, Woodcroft BJ et al (2014) GroopM: an automated tool for the recovery of population genomes from related metagenomes. PeerJ 2:e603

Iwai S, Weinmaier T, Schmidt BL et al (2016) Piphillin: improved prediction of metagenomic content by direct inference from human microbiomes. PLoS One 11:e0166104

Janda JM, Abbott SL (2007) 16S rRNA gene sequencing for bacterial identification in the diagnostic laboratory: pluses, perils, and pitfalls. J Clin Microbiol 45:2761–2764

Ji P, Zhang Y, Wang J, Zhao F (2017) MetaSort untangles metagenome assembly by reducing microbial community complexity. Nat Commun 8:1–14

Jünemann S, Kleinbölting N, Jaenicke S et al (2017) Bioinformatics for NGS-based metagenomics and the application to biogas research. J Biotechnol 261:10–23

Kanehisa M, Sato Y, Kawashima M et al (2016a) KEGG as a reference resource for gene and protein annotation. Nucleic Acids Res 44:D457–D462

Kanehisa M, Sato Y, Morishima K (2016b) BlastKOALA and GhostKOALA: KEGG tools for functional characterization of genome and metagenome sequences. J Mol Biol 428:726–731

Kang DD, Froula J, Egan R, Wang Z (2015) MetaBAT, an efficient tool for accurately reconstructing single genomes from complex microbial communities. PeerJ 3:e1165

Kang DD, Li F, Kirton E et al (2019) MetaBAT 2: an adaptive binning algorithm for robust and efficient genome reconstruction from metagenome assemblies. PeerJ 7:e7359

Kelley DR, Liu B, Delcher AL et al (2012) Gene prediction with glimmer for metagenomic sequences augmented by classification and clustering. Nucleic Acids Res 40:e9–e9

Konstantinidis KT, Stackebrandt E (2013) Defining taxonomic ranks. Prokaryotes 1:229–254

Konstantinidis KT, Tiedje JM (2007) Prokaryotic taxonomy and phylogeny in the genomic era: advancements and challenges ahead. Curr Opin Microbiol 10:504–509

Krause L, Diaz NN, Goesmann A et al (2008) Phylogenetic classification of short environmental DNA fragments. Nucleic Acids Res 36:2230–2239

Kristiansson E, Hugenholtz P, Dalevi D (2009) ShotgunFunctionalizeR: an R-package for functional comparison of metagenomes. Bioinformatics 25:2737–2738

Kuczynski J, Lauber CL, Walters WA et al (2012) Experimental and analytical tools for studying the human microbiome. Nat Rev Genet 13:47–58

Kultima JR, Sunagawa S, Li J et al (2012) MOCAT: a metagenomics assembly and gene prediction toolkit. PLoS One 7:e47656

Langille MGI, Zaneveld J, Caporaso JG et al (2013) Predictive functional profiling of microbial communities using 16S rRNA marker gene sequences. Nat Biotechnol 31:814

Langmead B, Salzberg SL (2012) Fast gapped-read alignment with bowtie 2. Nat Methods 9:357
Li H, Durbin R (2010) Fast and accurate long-read alignment with Burrows–Wheeler transform. Bioinformatics 26:589–595
Li W, Godzik A (2006) Cd-hit: a fast program for clustering and comparing large sets of protein or nucleotide sequences. Bioinformatics 22:1658–1659
Li F, Song J, Yang H et al (2009) One-step synthesis of graphene/SnO2 nanocomposites and its application in electrochemical supercapacitors. Nanotechnology 20:455602
Li D, Luo R, Liu C-M et al (2016) MEGAHIT v1. 0: a fast and scalable metagenome assembler driven by advanced methodologies and community practices. Methods 102:3–11
Liu B, Pop M (2011) MetaPath: identifying differentially abundant metabolic pathways in metagenomic datasets. In: BMC proceedings. BioMed Central, pp 1–12
Liu Y, Guo J, Hu G, Zhu H (2013) Gene prediction in metagenomic fragments based on the SVM algorithm. BMC Bioinform 14:S12
Luo C, Tsementzi D, Kyrpides NC, Konstantinidis KT (2012a) Individual genome assembly from complex community short-read metagenomic datasets. ISME J 6:898–901
Luo C, Tsementzi D, Kyrpides N et al (2012b) Direct comparisons of Illumina vs. Roche 454 sequencing technologies on the same microbial community DNA sample. PLoS One 7: e30087
Luo C, Rodriguez-R LM, Konstantinidis KT (2013) A user's guide to quantitative and comparative analysis of metagenomic datasets. Methods Enzymol 531:525–547
Luo C, Rodriguez-r LM, Konstantinidis KT (2014) MyTaxa: an advanced taxonomic classifier for genomic and metagenomic sequences. Nucleic Acids Res 42:e73–e73
Mallawaarachchi V, Wickramarachchi A, Lin Y (2020) GraphBin: refined binning of metagenomic contigs using assembly graphs. Bioinformatics 36(11):3307–3313
Mangrola AV, Dudhagara P, Koringa P et al (2015) Shotgun metagenomic sequencing based microbial diversity assessment of Lasundra hot spring, India. Genom Data 4:73–75
Markowitz VM, Chen I-MA, Chu K et al (2012) IMG/M: the integrated metagenome data management and comparative analysis system. Nucleic Acids Res 40:D123–D129
Martin JA, Wang Z (2011) Next-generation transcriptome assembly. Nat Rev Genet 12:671–682
McHardy AC, Martín HG, Tsirigos A et al (2007) Accurate phylogenetic classification of variable-length DNA fragments. Nat Methods 4:63–72
McKenna A, Hanna M, Banks E et al (2010) The genome analysis toolkit: a MapReduce framework for analyzing next-generation DNA sequencing data. Genome Res 20:1297–1303
McMurdie PJ, Holmes S (2015) Shiny-phyloseq: web application for interactive microbiome analysis with provenance tracking. Bioinformatics 31:282–283
Mende DR, Waller AS, Sunagawa S et al (2012) Assessment of metagenomic assembly using simulated next generation sequencing data. PLoS One 7:e31386
Meyer F, Paarmann D, D'Souza M et al (2008) The metagenomics RAST server–a public resource for the automatic phylogenetic and functional analysis of metagenomes. BMC Bioinform 9:386
Mikheenko A, Saveliev V, Gurevich A (2016) MetaQUAST: evaluation of metagenome assemblies. Bioinformatics 32:1088–1090
Mineeva O, Rojas-Carulla M, Ley RE et al (2020) DeepMAsED: evaluating the quality of metagenomic assemblies. Bioinformatics 36(10):3011–3017
Minot SS, Krumm N, Greenfield NB (2015) One codex: a sensitive and accurate data platform for genomic microbial identification. BioRxiv 27607. https://doi.org/10.1101/027607
Nagarajan N, Pop M (2013) Sequence assembly demystified. Nat Rev Genet 14:157–167
Namiki T, Hachiya T, Tanaka H, Sakakibara Y (2012) MetaVelvet: an extension of velvet assembler to de novo metagenome assembly from short sequence reads. Nucleic Acids Res 40:e155–e155
Noguchi H, Park J, Takagi T (2006) MetaGene: prokaryotic gene finding from environmental genome shotgun sequences. Nucleic Acids Res 34:5623–5630

Oulas A, Pavloudi C, Polymenakou P et al (2015) Metagenomics: tools and insights for analyzing next-generation sequencing data derived from biodiversity studies. Bioinform Biol Insights 9: BBI-S12462

Ounit R, Wanamaker S, Close TJ, Lonardi S (2015) CLARK: fast and accurate classification of metagenomic and genomic sequences using discriminative k-mers. BMC Genomics 16:236

Pace NR (1997) A molecular view of microbial diversity and the biosphere. Science 276:734–740

Parks DH, Imelfort M, Skennerton CT et al (2015) CheckM: assessing the quality of microbial genomes recovered from isolates, single cells, and metagenomes. Genome Res 25:1043–1055

Patel RK, Jain M (2012) NGS QC toolkit: a toolkit for quality control of next generation sequencing data. PLoS One 7:e30619

Patel R, Mevada V, Prajapati D et al (2015) Metagenomic sequence of saline desert microbiota from wild ass sanctuary, little Rann of Kutch, Gujarat, India. Genom Data 3:137–139

Paulson JN, Stine OC, Bravo HC, Pop M (2013) Differential abundance analysis for microbial marker-gene surveys. Nat Methods 10:1200

Peng Y, Leung HCM, Yiu S-M, Chin FYL (2012) IDBA-UD: a de novo assembler for single-cell and metagenomic sequencing data with highly uneven depth. Bioinformatics 28:1420–1428

Piro VC, Lindner MS, Renard BY (2016) DUDes: a top-down taxonomic profiler for metagenomics. Bioinformatics 32:2272–2280

Plaza Oñate F, Le Chatelier E, Almeida M et al (2019) MSPminer: abundance-based reconstitution of microbial pan-genomes from shotgun metagenomic data. Bioinformatics 35:1544–1552

Pop M, Phillippy A, Delcher AL, Salzberg SL (2004) Comparative genome assembly. Brief Bioinform 5:237–248

Poretsky R, Rodriguez-R LM, Luo C et al (2014) Strengths and limitations of 16S rRNA gene amplicon sequencing in revealing temporal microbial community dynamics. PLoS One 9: e93827

Prifti E, Zucker J-D (2015) The new science of metagenomics and the challenges of its use in both developed and developing countries. In: Socio-ecological dimensions of infectious diseases in Southeast Asia. Springer, Singapore, pp 191–216

Qian J, Comin M (2019) MetaCon: unsupervised clustering of metagenomic contigs with probabilistic k-mers statistics and coverage. BMC Bioinform 20:1–12

Ranjan R, Rani A, Metwally A et al (2016) Analysis of the microbiome: advantages of whole genome shotgun versus 16S amplicon sequencing. Biochem Biophys Res Commun 469:967–977

Rho M, Tang H, Ye Y (2010) FragGeneScan: predicting genes in short and error-prone reads. Nucleic Acids Res 38:e191–e191

Scholz MB, Lo C-C, Chain PSG (2012) Next generation sequencing and bioinformatic bottlenecks: the current state of metagenomic data analysis. Curr Opin Biotechnol 23:9–15

Shi Y, Tyson GW, Eppley JM, DeLong EF (2011) Integrated metatranscriptomic and metagenomic analyses of stratified microbial assemblages in the open ocean. ISME J 5:999–1013

Simon C, Daniel R (2010) Construction of small-insert and large-insert metagenomic libraries. In: Metagenomics. Humana Press, Totowa, NJ, pp 39–50

Sims D, Sudbery I, Ilott NE et al (2014) Sequencing depth and coverage: key considerations in genomic analyses. Nat Rev Genet 15:121

Singh AH, Doerks T, Letunic I et al (2009) Discovering functional novelty in metagenomes: examples from light-mediated processes. J Bacteriol 191:32–41

Steele HL, Jaeger K-E, Daniel R, Streit WR (2009) Advances in recovery of novel biocatalysts from metagenomes. J Mol Microbiol Biotechnol 16:25–37

Su X, Pan W, Song B et al (2014) Parallel-META 2.0: enhanced metagenomic data analysis with functional annotation, high performance computing and advanced visualization. PLoS One 9: e89323

Tatusov RL, Galperin MY, Natale DA, Koonin EV (2000) The COG database: a tool for genome-scale analysis of protein functions and evolution. Nucleic Acids Res 28:33–36

Teeling H, Waldmann J, Lombardot T et al (2004) TETRA: a web-service and a stand-alone program for the analysis and comparison of tetranucleotide usage patterns in DNA sequences. BMC Bioinform 5:163

Teeling H, Fuchs BM, Becher D et al (2012) Substrate-controlled succession of marine bacterioplankton populations induced by a phytoplankton bloom. Science 336:608–611

Tringe SG, Von Mering C, Kobayashi A et al (2005) Comparative metagenomics of microbial communities. Science 308:554–557

Truong DT, Franzosa EA, Tickle TL et al (2015) MetaPhlAn2 for enhanced metagenomic taxonomic profiling. Nat Methods 12:902–903

Turnbaugh PJ, Gordon JI (2008) An invitation to the marriage of metagenomics and metabolomics. Cell 134:708–713

Wang G-Y-S, Graziani E, Waters B et al (2000) Novel natural products from soil DNA libraries in a streptomycete host. Org Lett 2:2401–2404

Wang Y, Leung HCM, Yiu SM, Chin FYL (2014) MetaCluster-TA: taxonomic annotation for metagenomic data based on assembly-assisted binning. BMC Genomics 15:S12

Wilkening J, Wilke A, Desai N, Meyer F (2009) Using clouds for metagenomics: a case study. In: 2009 IEEE international conference on cluster computing and workshops. IEEE, Piscataway, NJ, pp 1–6

Woese CR, Fox GE (1977) Phylogenetic structure of the prokaryotic domain: the primary kingdoms. Proc Natl Acad Sci 74:5088–5090

Wood DE, Lu J, Langmead B (2019) Improved metagenomic analysis with Kraken 2. Genome Biol 20:257

Wu M, Eisen JA (2008) A simple, fast, and accurate method of phylogenomic inference. Genome Biol 9:R151

Wu M, Scott AJ (2012) Phylogenomic analysis of bacterial and archaeal sequences with AMPHORA2. Bioinformatics 28:1033–1034

Wu Y-W, Simmons BA, Singer SW (2016) MaxBin 2.0: an automated binning algorithm to recover genomes from multiple metagenomic datasets. Bioinformatics 32:605–607

Yan Q, Wi YM, Thoendel MJ et al (2019) Evaluation of the CosmosID bioinformatics platform for prosthetic joint-associated sonicate fluid shotgun metagenomic data analysis. J Clin Microbiol 57:e01182

Ye Y, Doak TG (2009) A parsimony approach to biological pathway reconstruction/inference for genomes and metagenomes. PLoS Comput Biol 5:e1000465

Ye Y, Tang H (2009) An ORFome assembly approach to metagenomics sequences analysis. J Bioinforma Comput Biol 7:455–471

Zhou Q, Su X, Jing G, Ning K (2014) Meta-QC-Chain: comprehensive and fast quality control method for metagenomic data. Genomics Proteomics Bioinformatics 12:52–56

Zhu W, Lomsadze A, Borodovsky M (2010) Ab initio gene identification in metagenomic sequences. Nucleic Acids Res 38:e132–e132

The manufacturer's authorised representative in the EU is Springer Nature Customer Service Centre GmbH, Europaplatz 3, 69115 Heidelberg, Germany. If you have any concerns regarding our products, please contact ProductSafety@springernature.com

Printed and bound by CPI Group (UK) Ltd, Croydon, CR0 4YY
30/09/2026
02225317-0004